Engineering Materials

Carolin Koerner

Integral Foam Molding of Light Metals

Technology, Foam Physics and Foam Simulation

With 143 Figures and 8 Tables

Springer

Dr. Carolin Koerner
Univ. Erlangen-Nürnberg
Inst. Werkstoffwissenschaften
Martensstr. 5
91058 Erlangen
Germany

ISBN: 978-3-540-68838-9 e-ISBN: 978-3-540-68839-6

Engineering Materials ISSN: 1612-1317

Library of Congress Control Number: 2008929338

Cover Design: Mönnich, Max

Printed on acid-free paper

9 8 7 6 5 4 3 2 1

springer.com

For Viktor and Walter

Preface

A person with a new idea is a crank until the idea succeeds.

Mark Twain

Metal foams show outstanding properties: Low weight, high rigidity, high energy absorption capacity, high damping capacity, etc. They have attracted strong industrial and scientific interest during the last decade. A variety of methods has been developed to produce foams and the development of new, more sophisticated methods is still going on.

On the one hand, there are only very few applications where metal foams can be directly employed without further processing. On the other hand, established metal foam production methods have one feature in common, they produce foam and not metal parts containing metal foam. In the majority of cases additional shaping and joining steps are necessary to transform the metal foam into a working functional element. In addition, the cellular structure demands for appropriate joining technologies which are often not yet available or expensive. As a result, the whole processing sequence is in general long and expensive.

The logical consequence of the requirement to develop cost-effective techniques to produce metal parts with integrated cellular structure is the newly developed process of integral foam molding. Integral foam consists of a solid skin and a cellular core. This is the fundamental construction principle which is ubiquitous in biological systems, e. g. the human skull, as well as in technical solutions, e. g. sandwich constructions. The concentration of the material within the skin optimizes the moment of inertia and thus stiffness and strength.

The development of metal based integral foam follows analogous paths as that of polymers where integral foam was commercially introduced in the late 1960s. Polymer integral foam parts are now accepted as a material system with characteristic properties which simplifies designs, reduces production costs and weight, and which increases stiffness and overall strength. Starting from low-pressure injection of polymers with small amounts of gas content, highly sophisticated processes have been developed where injection takes place at very high pressures into mold cavities with moving elements.

Metal integral foams represent – analogous to polymers – an own class of structures with specific properties such as high structural rigidity, high energy absorption capacity or high damping. They are not thought to substitute standard die castings but open new applications and also demand a specific component design. Compared to polymers, the development of metal integral foam molding is just at the beginning. The time delay of four decades between commercialization of polymer integral foam molding and the current onset of investigations with respect to metals can hardly be explained. Although we have made enormous progress during the last years of metal integral foam molding research there are still many challenges left which have to be solved in the future to get integral foam components made from light metals into broad industrial application. Nevertheless, we presume a large economical potential for metal integral foam molding technologies in the future.

Surprisingly, integral foam molding of metals shows much more similarities than differences to polymer integral foam molding. This is not only true for the applied molding technologies but also for the resultant integral foam structures and their properties. The intention of this book is to introduce the technological principles of metal integral foam molding. In addition, its purpose is to reveal the underlying physical mechanisms which govern the foam evolution process and which are essential in view of a successful process development.

Usually, a first step to gain a physical picture of the integral foam molding process would be to observe it. Unfortunately, in situ experimental observations are impossible since the foaming process takes place in a permanent steel mold within a fraction of a second. Thus, information of the foam formation process has to be extracted from ex situ investigations. This limited information does not help very much to understand the underlying physics. Nevertheless, in combination with theoretical investigations and numerical simulation the experimental findings reveal the underlying physical principles.

The book is organized in three parts:

- Part I: Technology
  Part I describes the technological foundation of integral foam molding of metals, a technology that has been conceived and developed in our research group at the University of Erlangen throughout the last 6 years. For the first time it is shown that foaming of metals is possible by applying molding techniques very similar to polymer integral foam molding. A low pressure and a high pressure integral foam molding process are introduced and discussed. The molded parts show compact skins and foamed cores with porosities up to 80%. Thermodynamics and kinetics of the blowing agent as well as the low viscosity of metal melts turn out to be the key for the success of the molding process. Although non-conditioned metal melts are employed, which are generally believed to be not foamable, the resulting structures are in fact foams. Integral foam molding appears to be the only known process where non-conditioned metal melts may be used to produce foam. An explanation of this finding is given on the basis of the theoretical background presented in Part II and Part

III. The fact that standard casting alloys may be applied has enormous advantages with respect to casting behavior, properties, recycling and, last but not least, costs.

- Part II: Physics
  This part is devoted to the physics of foaming with special emphasis on the very short time scale which is characteristic for integral foam molding. Although very complex in detail, foam formation is shown to underlie simple evolution laws determined by the way how foam stabilization is realized. Basis for these investigations is a numerical approach which is described in Part III. In order to account for the specific situation during integral foam molding, foam evolution in the presence of solid particles is discussed in detail. The results of these theoretical considerations represent the basis for the interpretation of the experimental observations described in Part I. However, the evolution laws are generally valid and can as well be applied to other foam evolution processes.

- Part III: Numerical Simulation
  The high degree of complexity of foam evolution processes strongly restricts analytical descriptions. Even with the aid of numerical tools, the simulation of foam formation processes is a challenge due to the huge and strongly evolving gas–liquid interface. A new lattice Boltzmann approach for the treatment of free surfaces is developed and applied on foam evolution problems. For the first time, the numerical simulation of foam evolution starting from nucleation until decay is accessible. The interplay between hydrodynamics, capillary forces, gravity and bubble coalescence processes leads to complex phenomena such as topological rearrangements, avalanches, drainage, etc. without further model assumptions.

Even if integral foam molding of light metals will never find general industrial application, we have gained fundamental insights into the main mechanisms of foam evolution, especially into the general principles of metal foam stabilization.

Acknowledgments
This work is based on results obtained during my work as the leader of the lightweight materials group at the Institute of Science and Technology of Metals at the University of Erlangen-Nuremberg.

I appreciate the support and freedom I obtained from Prof. R. F. Singer, the head of the Institute of Science and Technology of Metals, as it enabled me to follow-up and realize own ideas.

I would like to thank all members of the Institute for Materials Science who contributed to this work in one or another way. Although it is impossible to acknowledge all contributors, the work of the following colleagues is especially appreciated: M. Arnold, Dr. M. Hirschmann, Dr. M. Thies, A. Trepper and H. Wiehler.

In addition, I would like to thank Prof. Rüde, Prof. Kaptay, Prof. Greil and Prof. Clyne for the support during the habilitation process and the colleagues of Neue Materialien Fürth GmbH for the assistance with the thixomolder.

Carolin Körner, April 2008

Contents

Part I
TECHNOLOGY

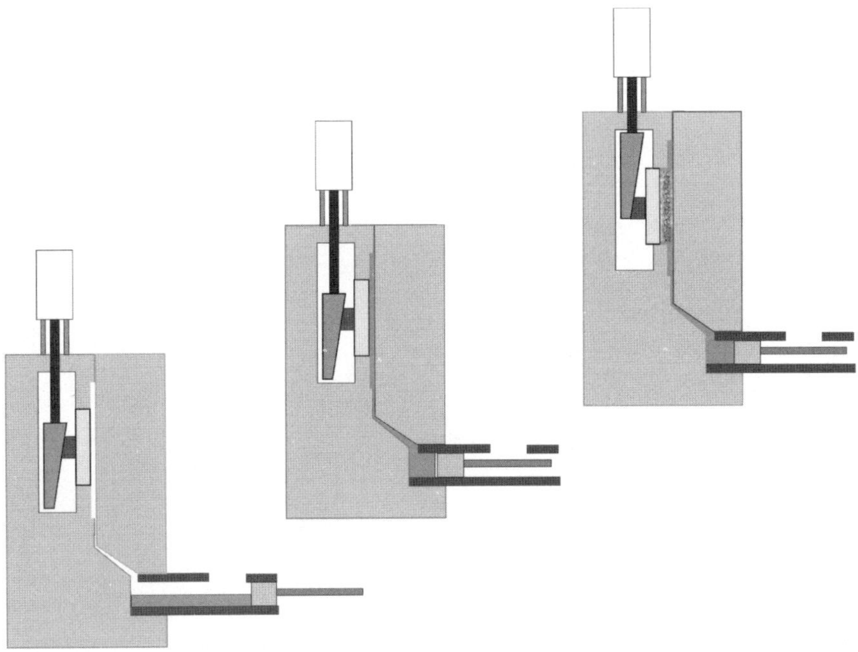

Es ist nicht genug, zu wissen, man muss auch anwenden; es ist nicht genug, zu wollen, man muss auch tun.

Johann Wolfgang von Goethe

Integral foam molding of light metals is based on very similar technologies developed and used in industry to produce polymer integral foams. We apply molding techniques – die casting and thixomolding – where liquid metals mixed with a blowing agent are injected with high velocities into permanent steel molds. The basic idea is that a compact layer develops at the mold surface due to the chilling effect of the cold mold wall. The solidification time of the kernel of the casting is much longer and the decomposition of the blowing agent leads to a foamed core.

When we started this development, there were some serious arguments why this general approach would be prone to fail. Firstly, the viscosity of metal melts is very low compared to polymers. Normally, a high viscosity is said to improve the stability of foams. Thus, the very low viscosity is expected not to allow the formation of a foamed core. Secondly, the solidification of metals proceeds much faster than that of polymers. Whether this extreme short time scale allows the decomposition of the blowing agent and the development of a cellular structure was questionable.

During process development it became more and more evident that the low viscosity is not a disadvantage of metals but might be turned into an advantage. There are two aspects why the low viscosity is a benefit. On the one hand, the low viscosity combined with the high mold filling velocities leads to turbulent melt flows which helps to admix the blowing agent into the melt in a homogeneous way. On the other hand, the mold filling behavior is completely different from that of polymers and supports the development of the solid surface layer. Also the very short solidification times have shown to be not only critical but also to be the reason why the solid skin is in fact compact and not the result of the compression of the surface bubbles of a former foam.

The blowing agent and its decomposition dynamics is the key for the success of the whole process. On the one hand, the decomposition kinetics has to be slow enough to prevent premature gas release. On the other hand, the decomposition has to be fast enough in order to guarantee that most of the blowing gas is released until solidification puts an end to the whole foaming process. Magnesium hydride meets these demands for light metals in a nearly perfect way and the resulting density profiles are the product of an interplay between gas release and solidification.

A very astonishing outcome of integral foam molding of light metals is that standard die casting alloys may be applied. The use of conditioned melts is not necessary in order to achieve a cellular foam core. This finding is quite surprising since it is well known that non-conditioned metal melts do not foam. Usually, particles have to be added to the melt to make metals foamable. The reason for this result is that foam formation takes place in the semi-solid state during integral foam molding. The solid phase particles show a barrier effect analogous to deliberately added ceramic particles. Thus, recycling of integral foam parts is also not an issue.

It is important to emphasize that integral foam molding is not thought to substitute standard die casting parts by foamed parts without changing the design. The appearance of integral foam parts is quite different from compact castings. Besides a strong weight reduction, they show interesting properties such as a high stiffness, high energy absorption capacity and strongly increased damping. Thus, integral foam parts open completely new applications and require a design adapted to process and application.

In summary, there are many similarities but also important differences between the processing of metals and polymers. The development of foam molding processes for metals, which is just at the beginning, may profit from the long experience available for polymers. Nevertheless, the specific properties of metal melts also open totally different process strategies.

1 Introduction

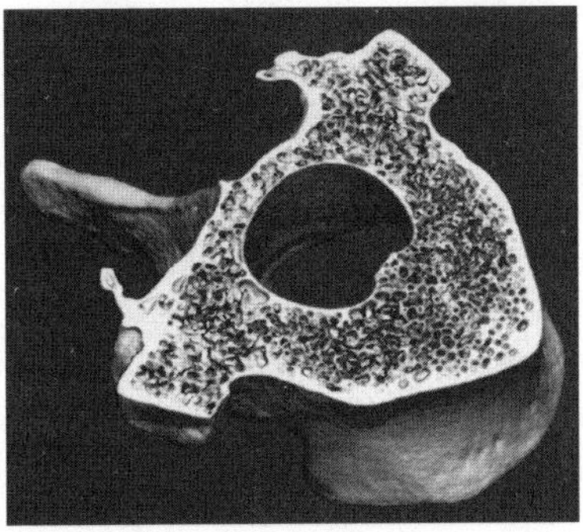

When modern man builds large load bearing structures, he uses dense solids: steel, concrete, glass. When nature does the same, she generally uses cellular materials: cork, wood, coral. There must be good reason for it.

Michael F. Ashby

1.1 Integral Foam

What is integral foam? The following sections are intended to introduce integral foam. We start with a short definition of the basic constituents of a foam structure. Subsequently, the advantages resulting from compact skins are briefly discussed and the notion *metal integral foam* is defined. The remainder of this chapter is devoted to the most important foam properties.

1.1.1 Foam

Foam develops by agglomeration of gas bubbles in a melt. The individual gas bubbles are hindered from assuming the energetically favorable round form due to the presence of neighboring bubbles. They are forced into a more or less polygonal shape, see Fig. 1.1.

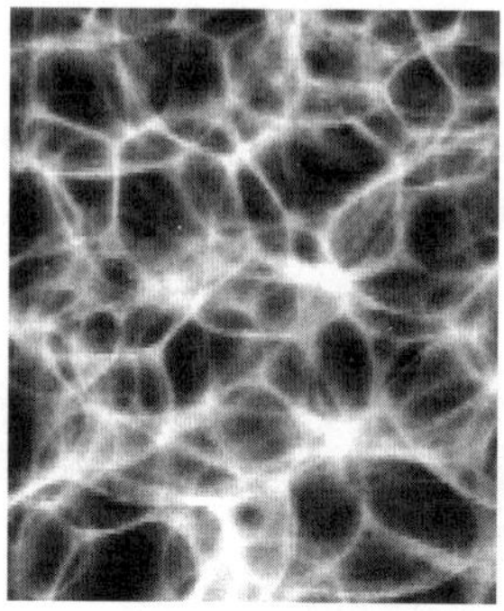

Figure 1.1: Polygonal foam structure: Radioscopy of an aluminum foam produced by foaming of a powder compact using TiH_2 as blowing agent [2, 17].

As constituents of a foam structure, gas bubbles are denoted as **cells**. The basic components of cells are **cell walls**, **Plateau borders** and **nodes**. A cell wall separates two gas bubbles over a length of about the bubble diameter and shows a curvature which is much smaller than the mean curvature of the two bubbles. Generally, the mean cell wall thickness is much smaller than the bubble diameter. The intersections of the walls are denoted as Plateau borders. The Plateau borders are interconnected in a disordered way and form a network whose nodes are junctions of at least four Plateau borders, see Fig. 1.2.

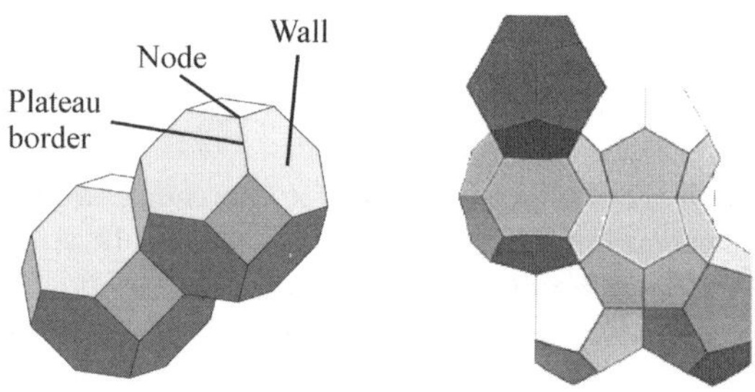

Figure 1.2: Idealized cells and their fundamental constituents. *Left*: Tetrakaidecahedra cell with walls, Plateau borders and nodes. *Right*: Weaire-Phelan cells [143].

The resulting cell geometry is determined by two conditions. On the one hand, the internal energy, i. e. the internal surface, should be as small as possible. On the other hand, the assembly of cells has to be space filling. The latter condition is a

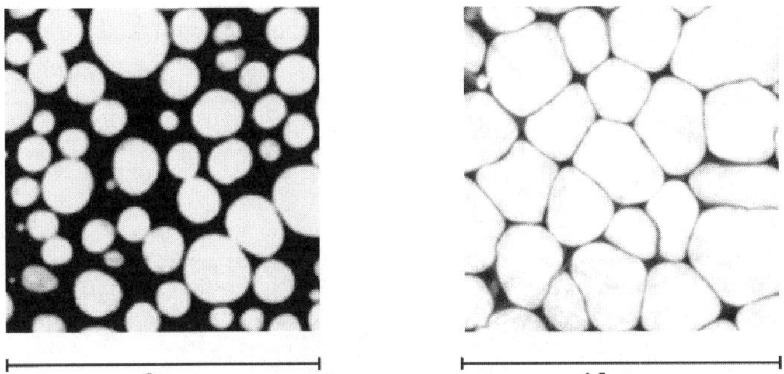

Figure 1.3: Wet (*left*) and dry (*right*) aluminum foam produced by foaming of a powder compact with TiH_2 as blowing agent [2, 17]. (μCT)

topological restriction and results from the fact that only a very limited number of cell geometries is actually space filling.[1] The search for the optimized cell is not yet finished. For a long time, the so-called Kelvin cell was believed to minimize energy. Today, the record is set by Weaire-Phelan cells [143] (see Fig. 1.2, right) and the search for even better structures is still going on.

Nevertheless, real foam structures are generally far away from the theoretical ones since they result from evolution processes which do not necessarily end up in optimal structures. A polygonal foam represents an ideal state which is often not or only to some extent achieved during production.

Generally, **dry foams** (polygonal foam) and **wet foams** are distinguished. The latter is also named as sphere foam since the cells are only slightly deformed without marked cell walls, see Fig. 1.3. A high liquid fraction allows the bubbles to be easily rearranged and makes the foam to behave like a liquid, i. e. the foam is not able to support shear forces. By successive draining out of the liquid due to gravity and capillary forces the wet foam eventually transforms into a dry foam with long, straight cell walls. In this state, the mobility of individual cells is extremely restricted and the foam is compelled to respond to outer forces by elastic cell deformation. Thus, dry foams are able to carry shear forces like ordinary solids.[2]

[1]This topological problem was already studied by Euler. He showed that the number of nodes, cell walls, Plateau borders and cells for a large aggregate of cells is connected in a defined way expressed by Euler's Law [85].

[2]The transformation from wet to dry foam is a classical phase transition which can be characterized by the shear modulus as order parameter [58, 109, 143, 170, 231]. For a study of the rheological and optical properties of Gillette shaving cream, see [116].

Depending on the underlying material system, the cell wall thickness ranges from 10^{-9} m to 10^{-4} m indicating the presence of different foam stabilization mechanisms. The typical cell wall thickness of aqueous foam in the dry limit is about 10^{-9} m with nearly all liquid located in the Plateau borders, see Fig. 1.1. However, the appearance of non-aqueous foams in the dry regime, especially metal foams, is generally quite different from aqueous foams, see Fig. 1.4. The cell wall thickness

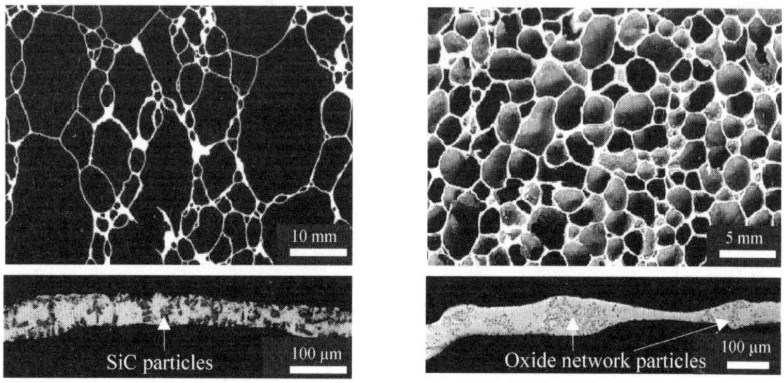

Figure 1.4: Aluminum foam. *Top*: Cell structures. *Bottom*: Cell walls. *Left*: Direct foamed SiC-particle reinforced aluminum melt by gas injection [181, 241]. *Right*: Foam produced by foaming of a powder compact with TiH_2 as blowing agent [2, 17]. In the following, these foams are named as PM foam.

is about 10^{-4} m and a substantial material fraction is located in the walls. The Plateau borders seem to be suppressed and sometimes they completely vanish. In addition, the cell shapes and arrangements appear to be far away from equilibrium. The latter observations are very important and reflect the underlying stabilization mechanisms.[3]

1.1.2 Integral Foam

The properties of foam can be significantly improved by surrounding it with compact skins. A compact skin has been shown to be indispensable for most technical applications of metal foam. Generally, a compact skin is demanded due to load transmission, stiffness and strength optimization or simply to prevent moisture from penetrating into the open foam structure. Although there is a variety of production

[3]We will come back to this point. Metal integral foams show similar structures.

methods available for metal foam [11], especially for aluminum, none of these methods allows the direct fabrication of foam parts with a sufficiently thick solid skin. Thus, foam is produced, e. g. by a powder metallurgical [2] or a melt route [181,241], and subsequently integrated into a component by encasing with a thick solid wall, e. g. by casting [32,108,144,148], roll-cladding [209], gluing, welding, thermal spraying, etc., see Fig. 1.5.[4] The structures in Fig. 1.5 have one thing in common – they

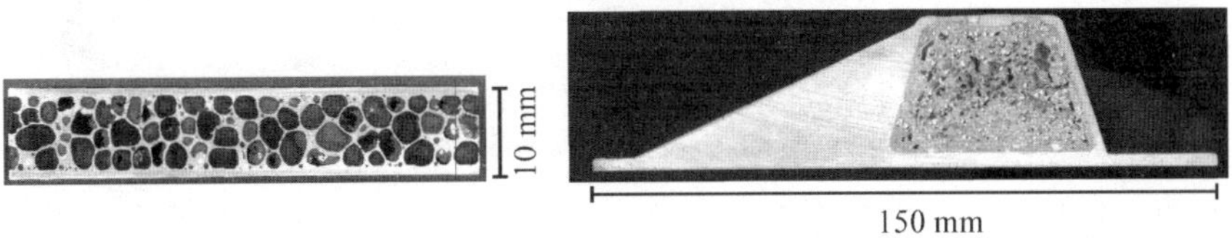

Figure 1.5: Integration of aluminum foam (PM foam) into a component. *Left*: Aluminum foam sandwich produced by roll-cladding (Source: Alm GmbH). *Right*: Aluminum foam encased with aluminum by die casting [148].

are not monolithic! Foam core and solid skin are produced in different processes. For the sandwich part (Fig. 1.5, left), conventional sheets of aluminum are roll-clad to a sheet of foamable precursor material. During heating, the foamable core expands while the roll-clad sheets remain compact. Alternatively, an aluminum foam part is produced and subsequently encased by casting (Fig. 1.5, right).

The processes developed to produce components including metal foam are – in most cases – expensive and show specific disadvantages. It is difficult to ensure a sound connection of the solid skin with the foam. Other drawbacks are restrictions with respect to the thickness of the compact skin. Encasing foam by sand-casting will process-related result in thick-walled parts.

Focusing our attention on polymers there exists a variety of molding or extrusion processes to produce foamed parts with a compact skin and a foamed core in one go. These kind of parts are denoted as integral foam.[5] Commonly, **polymer integral foam** is defined to be a molded or extruded part having a cellular core and an integral solid skin with the transition from the skin to the core being gradual [210]. We define **metal integral foam** in a fully analogous way. The main characteristics of integral foam are a compact skin with a gradual transition to a foamed core. This structure is a fundamental construction principle in biological systems e. g. the skeleton, see Fig. 1.6.

The skin gives the part its form and strength while the core can be used to optimize the weight-specific mechanical properties, e. g. the bending stiffness, the energy absorption or the damping properties.

[4]For details see *Handbook of Cellular Metals* [51].

[5]Integral foam is also designated as structural foam.

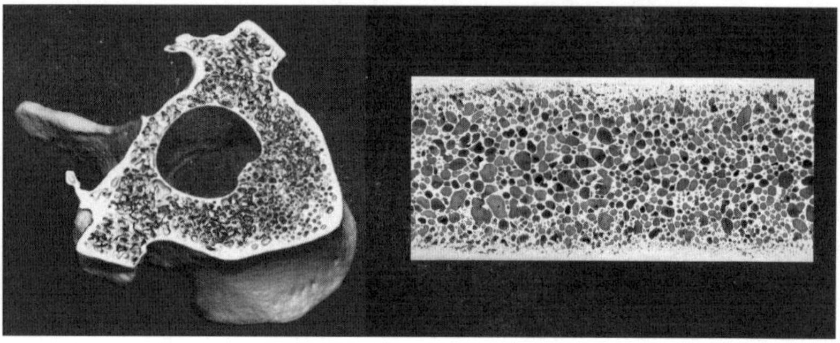

Figure 1.6: Integral foam. *Left*: Human vertebra as an example for an integral foam structure in nature. (Courtesy of Scanco Medical, 3D-μCT-reconstruction). *Right*: Aluminum integral foam (HP-IFM).

1.1.3 Foam Properties

Metal foams have attracted strong industrial and scientific interest during the last decade [156][6] since they possess outstanding mechanical properties: low weight, high specific bending stiffness,[7] high energy absorption capacity [103], high damping capacity, etc. [67]. These remarkable properties are an outcome of the cellular structure. Comprehensive descriptions of the mechanical properties of cellular materials can be found in the textbooks of Gibson and Ashby [85] and Hilyard [111]. In this section, the focus is on the most important aspects.

The quantity with the strongest impact on the physical and mechanical properties of solid foams[8] is the **relative density** ρ_{rel},

$$\rho_{\mathrm{rel}} = \frac{\rho_{\mathrm{F}}}{\rho}, \tag{1.1}$$

with the foam density ρ_{F} and the material density ρ.

The relative density and the **porosity** p are correlated by

$$p = 1 - \rho_{\mathrm{rel}}. \tag{1.2}$$

[6]Comprehensive overviews can be found in the reviews of Banhart [11], Ashby et al. [5] and Gibson [84].

[7]The specific elastic modulus of a foamed material is generally lower than that of the compact material.

[8]Solid foams develop from liquid foams by freezing of the structure due to solidification.

The **phase ratio** ϕ is defined as[9]

$$\phi := \frac{V_G}{V_F} = \frac{1}{\rho_{rel}} - 1 = \frac{p}{\rho_{rel}} \tag{1.3}$$

where V_G and V_F denote the total gas and fluid volume ($\equiv$ solid material volume for solid foams), respectively.

The compression behavior of foam is completely different from that of compact material due to its high compressibility [54, 85, 111]. A stress–strain curve for an ideal foam is depicted in Fig. 1.7. After a short elastic regime, foam starts to get compressed by successive collapse of cells as long as there are still intact cells present. During this regime – the stress plateau – the stress level is constant or only slightly increasing. As soon as there are no more intact cells present, the stress level starts to increase strongly due to compaction. This compression behavior is interesting with respect to energy absorption applications. A high amount of energy is absorbed on a comparatively low stress level [100].

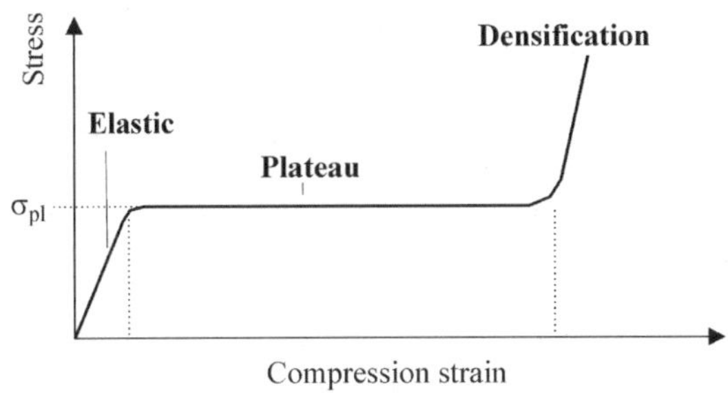

Figure 1.7: Schematic of the compression behavior of an ideal foam. σ_{pl} denotes the compression strength [85].

Many foam properties like the elastic modulus E_F, the compression strength σ_{pl}, the thermal conductivity λ_F, etc., follow a power law with respect to the relative density. The simplest relation is that between ρ_{rel} and the thermal or electrical conductivity [10, 85],

$$\lambda_F = \lambda_M \cdot C \cdot \rho_{rel}, \tag{1.4}$$

where the constant C accounts for the internal foam structure and λ_F and λ_M are the foam and material conductivity, respectively.

[9] Physical and mechanical properties of foams are governed by the relative density ρ_{rel}. On the other hand, foam evolution is controlled by the phase ratio, see Chap. 5.

The situation gets a little bit more complicated if different mechanisms have to be taken into consideration as it is the case for the mechanical properties. The foam stiffness E_{F} and the compression strength σ_{pl} are determined by the different deformation behavior of the cell walls and the Plateau borders. The main deformation mechanism of Plateau borders is bending whereas tension and compression prevail in cell walls. The elastic modulus E_{F} and the compression strength σ_{pl} are reasonably well approximated by [85],

$$E_{\mathrm{F}} = E\left[C_1\,\Phi^2\,\rho_{\mathrm{rel}}^2 + C_2\,(1-\Phi)\,\rho_{\mathrm{rel}}\right] \tag{1.5}$$

$$\sigma_{\mathrm{pl}} = \sigma_{\mathrm{y}}\left[C_3\,\Phi\,\rho_{\mathrm{rel}}^{\frac{3}{2}} + C_4\,(1-\Phi)\,\rho_{\mathrm{rel}}\right], \tag{1.6}$$

where $C_1, \ldots, C_4$ are constants which have to be experimentally determined. σ_{y} designates the yield strength of the solid material. The volume fraction of material contained in the Plateau borders is denoted by Φ with $0 \leq \Phi \leq 1$. The two limits, $\Phi = 0$ and $\Phi = 1$, represent the extremes where all material is concentrated in the cell walls and in the Plateau borders, respectively. The first term on the right hand side describes the behavior of the Plateau borders whereas the second one considers the cell walls. The different contributions are weighted by Φ in order to account for the material distribution between borders and walls. Although very simple, Eq. (1.5) and Eq. (1.6) have shown to be capable to describe the essential foam behavior.

As a general rule, inserting holes into a material – this is exactly what happens during foaming – worsens stiffness and strength. For what reason then should we use foam? Foam exhibits new properties such as a high specific bending stiffness and strength and a high energy absorption capacity. For certain loading cases, very stiff structures can be realized with foam [51, 85].

The damping properties of metal foams become more and more apparent as a very promising feature for noise or vibration damping applications [51, 189]. Although metal foam is said to show a high damping capacity, there are only few systematic investigations with respect to the underlying damping mechanisms [14, 88, 91, 92, 93, 94, 97, 101, 171].

The damping properties of metal foams are influenced by structural factors such as relative density, cell diameter and the base material as well as the testing conditions determined by the strain amplitude, the frequency, the temperature, etc. In addition, the damping characteristics are only stable for low amplitudes and exhibit instabilities at higher ones [93, 94].

Energy dissipation in a vibrating system causes a decrease of the vibration amplitude and thereby damping. Generally, damping is measured by determination of the decreasing amplitude of a free vibration,

$$A(t) = A_0 \exp\left(-\pi\,Q^{-1}\,f\,t\right), \tag{1.7}$$

where A is the time dependent vibration amplitude with $A_0 = A(0)$, f is the frequency and Q^{-1} denotes the internal friction.

Usually, various damping mechanisms are superimposed and different mechanisms are dominant in different amplitude or frequency ranges. Amplitude independent and amplitude dependent damping is distinguished. Generally, the internal friction may be represented as a sum of a strain independent, Q_0^{-1}, and a strain dependent, Q_ϵ^{-1}, contribution:

$$Q^{-1} = Q_0^{-1} + Q_\epsilon^{-1}. \tag{1.8}$$

The strain independent contribution is mainly due to reversible dislocation movement and thermoelastic internal friction whereas the strain dependent contribution is based on irreversible dislocation movement. Thus, Q^{-1} is strain independent for low strains and shows a strong increase at higher strains. Damping at higher strains, which might be rather high, eventually leads to fatigue related instabilities which has to be taken into account in case of industrial applications [93, 94].

The strain independent contribution may be represented as a sum [88]

$$Q_0^{-1} = Q_{0,\mathrm{b}}^{-1} + Q_{0,\mathrm{th}}^{-1}(f, d) \tag{1.9}$$

where $Q_{0,\mathrm{b}}^{-1}$ denotes the background damping (e. g. reversible dislocation bowing) and $Q_{0,\mathrm{th}}^{-1}$ comprises the thermoelastic effect. Material heats under compressive stress and cools under tensile stress. The thermoelastic effect describes damping owing to transversal heat flow in bending beams with thickness d from compressed to expanded regions at frequency f. Thermoelastic damping may reach high values for low frequencies and thin beams. For acoustic frequencies, the thermoelastic effect in compact materials with $d > 1\,\mathrm{mm}$ is rather small and proportional to $1/(f \cdot d^2)$ [92, 94]. Nevertheless, the relevant length scale for thermoelastic damping in metal foams might be determined by the cell diameter or the cell wall thickness rather than the thickness of the beam [88].

The fact that internal friction is increased in metal foam compared to compact material is generally agreed. In addition, it is also agreed that damping is related to the very complex inhomogeneous stress–strain state which develops in a foam subjected to cyclic loads. Nevertheless, the underlying mechanisms are still under discussion.

1.2 Polymer Integral Foam Molding

This section briefly describes the state-of-the-art of integral foam molding for polymers. Reviewing the fundamental principles of polymer integral foam molding is important since many principles can be directly transfered to metals. On the other hand, the specific differences between polymers and metals open also possibilities to develop completely new strategies.

Polymer integral foam molding is far more complex than conventional molding techniques and consists of three steps: admixing of a blowing agent to the polymer, filling of the mold cavity and, foaming the part in the cavity during cooling.

Historically, polymer integral foam molding developed from the effort to eliminate sink marks in thick wall moldings by addition of a small amount of blowing agent to the polymer. Afterward it was found that a larger amount of blowing agent leads to parts with a foam structure and a surface appearance similar to wood. Later on, a variety of different processes with specific advantages has been developed by raw material suppliers, machine manufacturers and technical institutes. For polymers, the high degree of complexity has led to substantial progress in process techniques during the last decades. A great number of different process strategies and technical realizations has been developed and is still under progress [18,35,59,128,182,187,216, 217,220]. In the following, the most important techniques are briefly described.[10]

1.2.1 Low Pressure Injection Molding of Polymers

The low pressure integral foam molding process has been known since the 60th and consists of the following steps [210]:

- Addition of a chemical or physical blowing agent[11] into the barrel.
- Plasticizing while maintaining the melt under a high pressure.
- Injection of the polymer containing the blowing agent into the cavity within a short time without filling it completely.
- Filling of the cavity by expansion of the blowing gas.

Form filling and foam evolution for the low pressure injection molding process is schematically depicted in Fig. 1.8.

Advantages of low pressure integral foams are a high specific rigidity, low specific weight, low internal stresses, no sink marks, good acoustic properties, etc. Shortcomings are the mechanical roughness and visible defects of the surface. Another drawback is the poor quality of extremities of molded parts. The main reason for these shortcomings is the relatively low dissociation pressure of at most 30 bar for chemical blowing agents [71].

[10] A comprehensive description can be found in [225] and in the textbook of Semerdjiev [210].

[11] Physical blowing agents are liquids or gases which get dissolved in the molten material by applying a high pressure. Chemical blowing agents are powder substances which release blowing gases when exposed to higher temperatures.

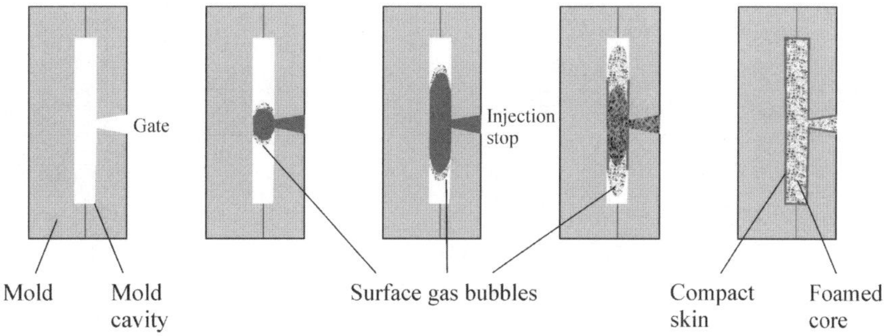

Figure 1.8: Low pressure integral foam molding process for polymers. The sudden pressure release during mold filling leads to surface bubble development and to a rough surface. The filling pressure is equal to the gas pressure of the blowing agent.

1.2.2 High Pressure Injection Molding of Polymers

The high pressure injection molding process was invented to overcome the inherent problems of the low pressure process. High pressure injection molding is much the same as conventional injection molding except that the volume of the cavity is initially smaller than that of the final part. The steps of the high pressure process are [183, 210]:

- Addition of a chemical or physical blowing agent into the barrel.
- Plasticizing while maintaining the melt under a high pressure.
- Injection of the polymer containing the blowing agent into the cavity within a short time (about 500 ms) and complete filling of the mold.
- Foaming of the part by expansion of the mold after a time delay of several seconds.

Form filling and foam evolution is depicted in Fig. 1.9. The surface quality is significantly improved compared to the low pressure process since surface bubbles get strongly compressed due to the high applied dwell pressure after the mold is completely filled. The surface quality is better but not perfect since the surface bubbles do not vanish but are only more strongly compressed. Besides an improved surface quality, lower densities can be realized compared to the low pressure process, see Fig. 1.10.[12] Drawbacks are a complex mold design due to moving cores or a vertical

[12]The problem of evolving surface bubbles is always present when the blowing agent is already dissolved in the melt and foaming is prevented by an applied pressure. When the pressure is removed bubbles develop virtually instantaneously.

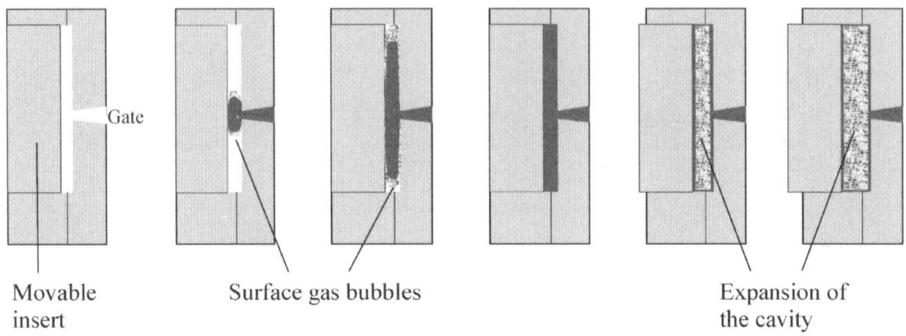

Figure 1.9: High pressure integral foam molding process for polymers. The sudden pressure release during mold filling leads to surface bubble development. These bubbles are compressed due to the dwell pressure. Foam develops due to the expansion of the completely filled cavity.

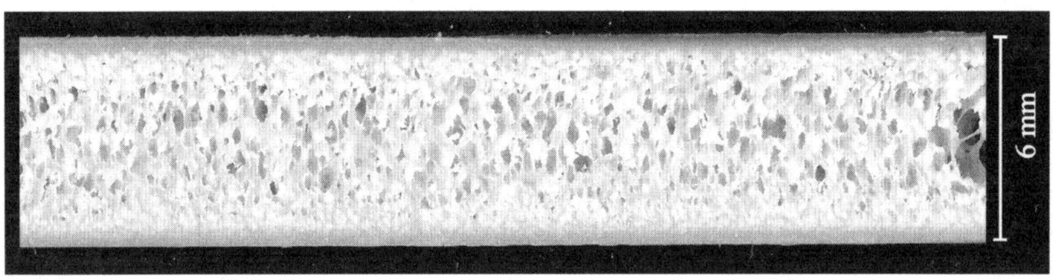

Figure 1.10: Structure of a high pressure thermoplastic foam molding (Polypropylene, Hostacom X 9067 HS, Basell, chemical blowing agent, Hydrocerol BM 40, 2.5 wt.%, courtesy of N. Müller, Chair of Polymer Technology LKT, University Erlangen – Nürnburg).

flash design and limitations to part geometry [59]. This method is best suited for simple flat geometries. A further inherent drawback are visible surface marks in those areas where mold expansion takes place.

A pronounced further improvement of the surface quality can be achieved by applying a counter pressure in the mold which suppresses premature bubble evolution at the filling front. Also additional heating of the mold surface or two-component processes are applied where the first component without blowing agent forms the surface and the second component with blowing agent the foam core [59].

1.2.3 MuCell™ Process

The MuCell™ process is based on physical blowing agents, especially nitrogen, and allows foaming pressures[13] of 100–200 bar. Originally, it was developed at the Massachusetts Institute of Technology (USA) and patented by Trexel Inc. (Woburn/USA). The MuCell™ process allows the production of injection moldings with a very fine cell structure with cell diameters between 1 and 100 μm [61, 62].[14] Thick and thin wall products can be produced. The essential characteristic of the process is that supercritical gases, N_2 or CO_2, are injected into the screw barrel at high pressures of 100–200 bar where they are homogeneously distributed in the polymer melt. The essential advantage of the MuCell™ process results from the much higher foaming pressures compared to chemical blowing agents. In order to maintain these high pressures and to achieve a homogeneous distribution of the gas in the melt special screw designs are necessary. Eventually, the gas has to be completely dissolved in the polymer melt. During injection, the sudden pressure drop and the high driving force for nucleation leads to a high number of tiny bubble nuclei. Depending on the application, weight reduction is between 10 and 30%.

The MuCell™ technology is the most frequently used technology today. The dissolved overcritical gas improves the flow characteristics of the polymer melt and the foaming pressure helps to fill thin sections. As a result, thin parts, less than 1 mm, can be produced. In addition, cycle time and clamping forces can be reduced. Normally, the MuCell™ technology is used in low pressure molding processes. However, lower overall densities can be realized by the high pressure method using breathing molds where the mold cavity is expanded by mold opening [177]. The breathing mold technique also shows that the development of tiny cells is only possible for high densities. For lower relative densities, $\rho_{rel} < 0.5$, the resulting cell diameter is much larger due to cell coalescence processes and similar to standard molding techniques [177].[15]

[13] We define foaming pressure as the gas pressure present during foaming, i. e. the pressure in the gas bubbles. During standard foaming processes, the foaming pressure is typically about 1 bar.

[14] Structures with cell sizes smaller than 100 μm are denoted as micro-cellular.

[15] The success of the MuCell™ technology is based on the high nuclei density. For higher porosities, the influence of the nuclei density vanishes, compare Chap. 5.

2 Integral Foam Molding of Metals

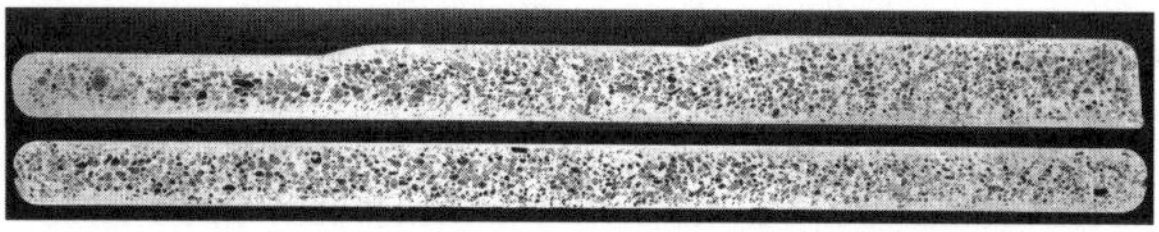

Is it possible to transfer processes developed for polymer injection molding to metals by applying similar casting technologies? There are many reasons why this is doubtful. The viscosity of molten metals is by a factor of 10^{-3}–10^{-6} smaller than that of polymer melts whereas the heat conductivity of metals is by a factor of 10^3 larger than that of polymers. Thus, solidification time is extremely short – for thin walled parts a fraction of a second. Whether bubble nucleation, decomposition of the blowing agent and foam expansion is possible during this short period is also not obvious. Moreover, a blowing agent has to be available which produces a sufficiently high foaming pressure.

The extremely low viscosity of metal melts may be the origin of additional difficulties. Even if bubble expansion is possible on this time scale, the low viscosity in combination with the high surface tension (about a factor of ten higher compared to polymers) is expected to provoke severe bubble coalescence and eventually the collapse of the whole cellular structure.[1]

In the following, it will become more and more evident that the just mentioned challenges turn out to be advantages of metals compared to polymers:

- A low viscosity combined with a high melt velocity results in turbulent melt flows for metals. Turbulences can be effectively used to admix the blowing agent to the liquid metal.
- The chilling effect of a cold mold wall combined with the high heat transfer rates leads to compact surface skins.
- The low viscosity of metals leads to a completely different form filling behavior compared to polymers.

[1]Perhaps these two arguments are the reason why metal integral foam molding has not been taken into consideration so far.

The main topic of this chapter is to describe the technological principles of metal integral foam molding. Several modifications of standard molding technologies are indispensable in order to produce integral foam parts. Firstly, admixing of a blowing agent has to be realized. Secondly, different mold designs and processing strategies – analogous to polymers – have to be developed.

2.1 Basic Considerations

2.1.1 Molding Technology

The development of metal based integral foam molding processes follows analogous paths as that of polymers. The underlying molding technology is die casting [33] or alternatively thixomolding [173], see Fig. 2.1.[2] The latter corresponds to injection molding for polymers. Magnesium is heated in a screw and generally processed in the semi-liquid state. The main difference with respect to die casting is that the viscosity of the semi-liquid material can be varied over several orders of magnitude.[3]

Die casting as well as thixomolding are characterized by a very high injection velocity of the molten metal. Aluminum as well as magnesium alloys may be processed by die casting whereas thixomolding is restricted to magnesium and can not be applied to aluminum alloys with the current machine technology.[4]

In both cases – die casting as well as thixomolding – the melt is injected into a permanent steel mold during the fraction of a second. Filling of the part cavity takes only about 20–100 ms. The velocity of the plunger during mold filling is about 1–5 m/s, thereby generating local melt velocities of about hundred m/s in the mold. These high velocities are essential for an efficient and homogeneous admixing of the blowing agent and for a good surface quality, see Sect. 2.2.[5]

The realization of an integral foam molding process for metals involves technical solutions for two challenges:

1. Admixing of the blowing agent (Sect. 2.2)

2. Development of adapted form fill processes and mold technologies (Sects. 2.3 and 2.4)

[2]The die casting experiments were accomplished on a Frech DAK 450-54 at the department of Materials Science, University Erlangen-Nürnberg. The thixomolding experiments were realized on a Thixomolding® machine from Japan Steel Works at NMF GmbH, Fürth.

[3]Nevertheless, the best foam molding results are gained in the fully liquid state.

[4]At temperatures of about 600°C the screw barrel would be strongly attacked by liquid aluminum.

[5]Below it will be explained how the high velocity is used to admix the blowing agent. Other casting methods like squeeze casting, investment casting or sand casting show a much lower melt velocity and are thus not suitable for our process.

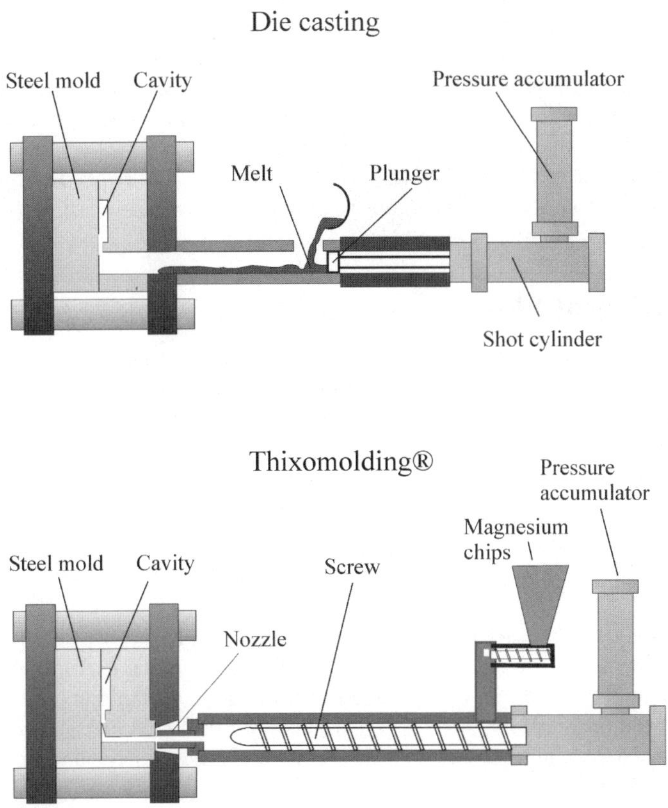

Figure 2.1: Basic methods: *Top*: Schematic sketch of a cold chamber die casting unit. *Bottom*: Schematic sketch of a thixomolding unit.

A direct transfer of successful polymer processes to metals without modifications is not possible due to the considerable property differences between metal and polymer melts. Nevertheless, the main principles may also be applied to metals. The main difference is the way of admixing the blowing agent. The extremely low viscosity of liquid metals proves to be essential for this process step.

During injection molding of polymer integral foam, the chemical blowing agent is admixed to the polymer granules and introduced into the molding machine. The polymer is plasticized under back pressure in the screw barrel of an injection-molding machine. The saturation pressure of the blowing gases can be regulated by controlling the pressure (5–30 bar) applied on the melt inside the injection cylinder [18, 210]. The gas remains dissolved as long as the applied pressure is high enough.

Is it possible and reasonable to employ an analogous process for metals, especially aluminum and magnesium? Prerequisite for an analogous approach is the availability of an analogous machine with a screw barrel, e. g. a thixomolding machine. It is important to note, that the involved pressures in the screw barrel of a thixomolding machine are low, i. e. the state-of-the-art does not allow to apply high pressures. Thus, an admixing procedure analogous to polymers is not supported by the present thixomolding technology and an alternative has to be found.

2.1.2 Turbulent Flow

The standard casting method for light metals is die casting [33] which is much simpler than injection molding since plasticizing is not an issue for metals. The screw barrel is replaced by a simple chamber, see Fig. 2.1. The metal is molten in a furnace, dosed into the chamber and directly injected into the mold with plunger velocities of 1–5 m/s. The local velocities v in the steel mold are about 10–100 m/s. The characteristic Reynolds numbers[6] Re are thus

$$Re = \frac{D\,v}{\nu} \approx 10^4 - 10^6, \tag{2.1}$$

where $D = 0.001$–0.01 m is the characteristic length and $\nu = 10^{-6}$ m^2/s the kinematic viscosity.[7] During die casting, the Reynolds number exceeds the critical Reynolds number, which describes the transition between laminar and turbulent flow, by several orders of magnitude. Thus, the flow during die casting shows turbulent character with the microscopic fluid particles moving in irregular, wavy paths, see Fig. 2.2.

2.1.3 Spray Filling

Not only the paths of the microscopic particles but also the macroscopic fluid behavior is strongly influenced by the melt velocity and its viscosity. During die casting, form filling is similar to spray filling where the melt front disintegrates into many tiny droplets. This is in contrast to polymer molding where form filling has laminar character due to the much higher viscosities, see Fig. 2.3.

Consequently, the large difference of the viscosities of polymer and metal melts results in a completely different form filling behavior. Polymer melts fill the mold in

[6] In 1883 Osborne Reynolds defined a dimensionless parameter, the Reynolds number, which indicates the transition from laminar to turbulent flow in pipes. For the flow inside a tube of diameter D he determined the critical value of this parameter to be 2300.

[7] This is about the kinematic viscosity of liquid light metals.

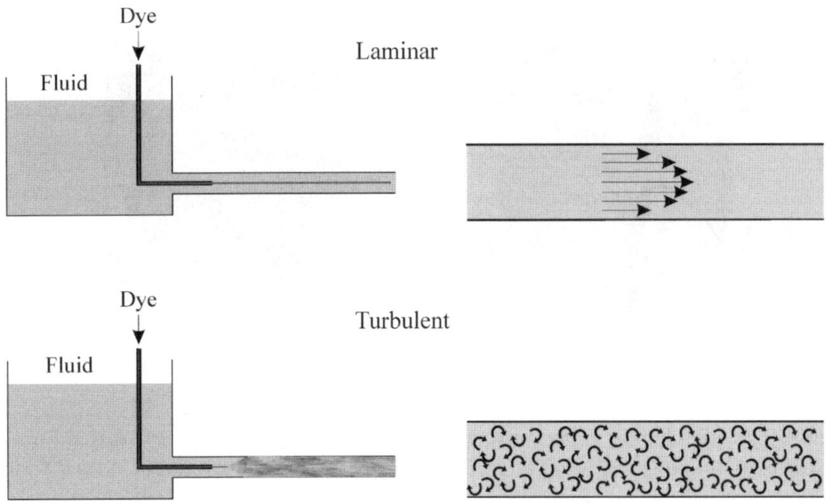

Figure 2.2: Schematic of the famous Reynolds experiment. *Top*: Laminar flow, *Bottom*: Turbulent flow. The eddies, which are characteristic for turbulent flows, destroy the flow lines and distribute the dye all over the pipe.

a laminar way with a closed melt front whereas metals are sprayed. As a result, during the initial phase of mold filling metals tend to wet the mold walls completely. During the later phase of mold filling, filling of the core of the casting takes place.

The mold filling behavior of metals, which is dominated by their low viscosity, has essential advantages compared to that of polymers:

- Admixing of the blowing agent
  Generally, mold filling for metals has to be regarded as turbulent. The kinetic energy of the metal melt during the form filling process leads to strong local mixing effects. It is exactly this behavior of metal melts which we employ for the admixing of the blowing agent.

- Foam formation
  The fronts of the foam are situated in the interior of the casting and the distances which have to be filled by foam expansion are short (Fig. 2.4, left). This behavior has essential advantages with respect to the resulting surface quality. If whole regions including the surface of the casting have to be filled by foam expansion – as it is the case for the low pressure integral foam molding

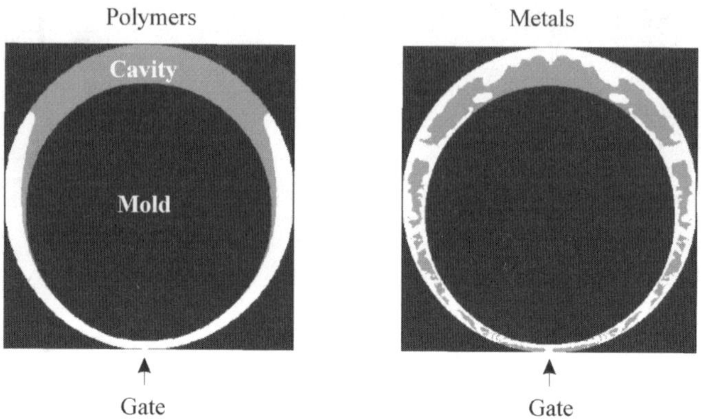

Figure 2.3: Illustration of the mold filling behavior of polymers (*left*) with $Re \approx 50$ compared to metals (*right*) with $Re \approx 1000$, which is representative for metal melts with a high solid phase fraction. (2D-LBM simulation)

process for polymers – the foam touches the mold surface and is subsequently compressed. As a result, the bubble structure of the foam is reproduced on the surface (Fig. 2.4, right).

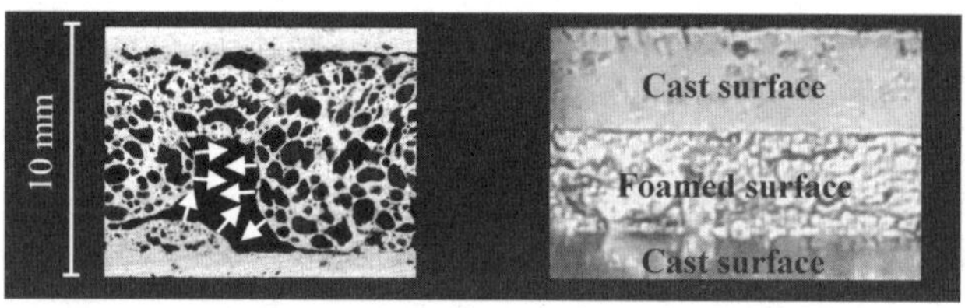

Figure 2.4: *Left*: Foam fronts in the interior of the casting. *Right*: Cast vs foamed surface. (high pressure integral foam molding, AlSi9Cu3)

2.2 Admixing of the Blowing Agent

The primary problem during integral foam molding is to admix the blowing agent to the molten metal. Unfortunately, the obvious approach to proceed in an analogous way as for polymers, which means to add the blowing agent in the screw barrel of a thixomolding machine, would be involved with enormous technical difficulties. However, the quite different fluid dynamical properties of metals compared to polymers open an interesting alternative.

2.2.1 Turbulent Mixing

Standard die casting processes show Reynolds numbers much higher than the critical one and are thus strongly turbulent. On the other hand, a fundamental property of turbulence[8] is its diffusivity which causes rapid mixing and increased rates of momentum, energy and mass transfer, see Fig. 2.2. For this reason, it is obvious to evaluate the possibility of turbulent mixing for the admixing of the chemical blowing agent.

In addition to admixing, a short contact time between blowing agent and melt or a high ambient pressure are important to prevent premature gas release and foaming. For polymers, where the contact times are about minutes, this problem is solved by applying a high pressure.[9] As already mentioned, employing a high pressure during a metal casting process is not practicable so far. Hence, the only option to prevent premature foam formation is to minimize the contact time.

Both requirements, turbulent mixing and a short contact time, can be realized by addition of the blowing agent into the melt flow during or just before melt injection. The key question is where the blowing agent should be added – in the injection cylinder, in the mold cavity or somewhere in between, see Fig. 2.5.

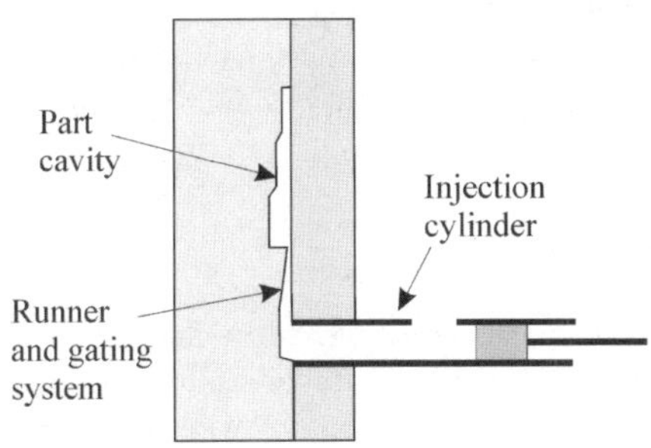

Figure 2.5: Die casting mold with injection cylinder, runner system and part cavity (schematic).

[8] Heisenberg reportedly said on his deathbed that he would ask God two questions: Why relativity and why turbulence? I (Heisenberg) really think he may have an answer to the first question.

[9] However, during mold filling the pressure diminishes at the filling front and local foaming occurs, see Sect. 1.2.

If the blowing agent is deposited in the part cavity, lumps of powder are smashed by the incoming melt to the mold walls or accumulate in shaded regions. Only a part of the blowing agent becomes actually thoroughly mixed with the melt. Adding the blowing agent in the part cavity is obviously not optimal.

One alternative would be the addition in the injection cylinder, e. g. during the dosing step. The danger of hydrogen explosion[10] can surely be solved by an adequate sealing of the chamber but there is still an efficiency problem since the blowing agent is present throughout the melt. Taking into account that in most cases about 50% of the weight of a die casted part belongs to the runner system it becomes clear that this procedure is not efficient. However, the main problem is not efficiency but premature gas release. The contact times between blowing agent and metal melt are such high that most of the blowing gas is already released in the casting chamber. The gas is either lost to the atmosphere or premature foaming starts. Both should be prevented.

The last option is the addition of the blowing agent in the runner and gating system located between injection cylinder and part cavity [151]. This method has various advantages:

- Short contact times < 100 ms.
- High Reynolds numbers when the melt gets in touch with the powder.
- Lumps of powder smashed on the wall are located in the runner system and not in the part.

Taking all arguments together, addition of the blowing agent in the runner system located between injection cylinder and part cavity appears the most practicable approach.[11] An ideal dosing system would be one working in a similar way as the dye introduction in Fig. 2.2.

2.2.2 Blowing Agent Feeding

Feeding of the blowing agent is accomplished by depositing the blowing agent into the runner system between casting chamber and component, see Fig. 2.6. One possibility to realize deposits is to envelope blowing agent powder by a foil, e. g. an aluminum foil. The metal melt has to be separated from the blowing agent during filling of the casting chamber in order to prevent premature decomposition. During the shot, the incoming melt catches the blowing agent and transports it to the gate. This approach has shown to be very effective since nearly all blowing agent powder

[10] Metal hydrides are typically used as blowing agents.

[11] An analogous proceedure is not practicable for polymers since the laminar flow behavior does not lead to an adequate mixture.

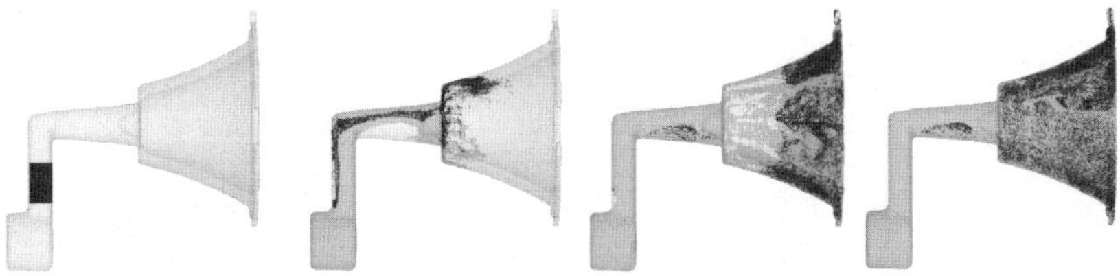

Figure 2.6: Feeding and distribution of the blowing agent (dark spots) for a bell-shaped component simulated with the commercial software Flow-3D.

reaches the mold cavity. In addition, the distribution of blowing agent particles is astonishingly homogeneous allover the casting.[12]

2.2.3 Amount of Blowing Agent

The amount of blowing agent[13] is about 0.5–2 wt.% of the integral foam part. This quantity of blowing agent is about the same as in other blowing agent based metal foaming methods [11, 181].

2.3 Low Pressure Integral Foam Molding

Low pressure integral foam molding (LP-IFM) for metals [114, 115, 151, 152, 153, 213][14] corresponds to that for polymers in Sect. 1.2.1. The liquid metal is injected into a permanent steel mold with velocities analogous to die casting conditions (plunger velocity about 1–5 m/s). In contrast to standard die casting, the mold cavity is not completely filled. Subsequently, the plunger may either be completely stopped or a very low dwell pressure[15] of about 5–10^2 bar may be applied, see Fig. 2.7.

[12]We have shown that this procedure works surprisingly well for parts up to 1 kg. Whether more sophisticated methods have to be developed for larger parts and larger amounts of blowing agents is unclear at the moment.

[13]The blowing agent is discussed in detail in Sect. 2.5.2.

[14]A similar process for aluminum is described in the more recent patent of Goldschmidt GmbH and Bühler Druckguss AG, see [190]. The only available publication is [25] where the process is briefly described. The blowing agent is added into the runner system with the help of a small plunger synchronized with the real time controlled plunger of a Bühler die casting machine. Experiments were only performed for aluminum. Material properties are not reported.

[15]The dwell pressure p_{dwell} denotes the driving pressure of the machine, i. e. the applied hydraulic pressure of the plunger. The dwell pressure is not identical to the casting pressure p_{cast} in the casting chamber acting on the liquid metal: $p_{\text{cast}} = p_{\text{dwell}} \cdot D_{\text{driv}}/D_{\text{cast}}$ where D_{driv} and D_{cast} denote the diameter of the driving plunger and the casting plunger, respectively.

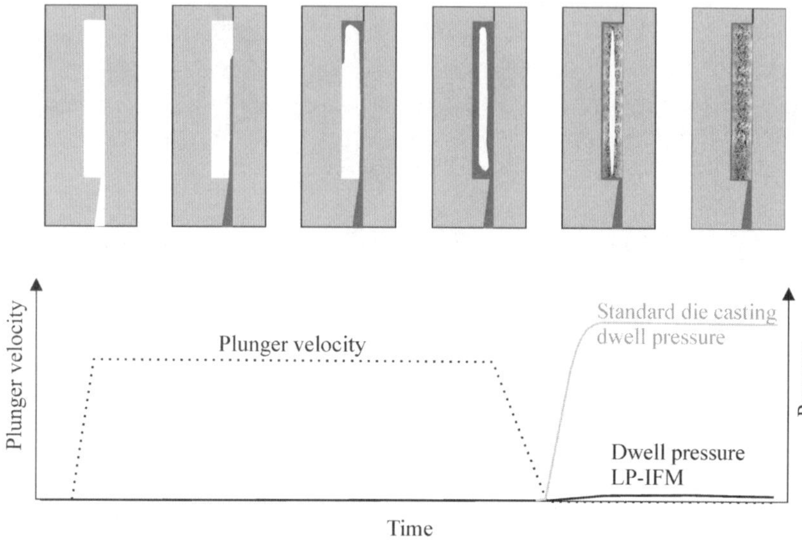

Figure 2.7: Low pressure integral foam molding of metals (schematic).

Stopping of the plunger can be realized with a standard real time controlled die casting machine without any modifications. When the plunger is frozen there is still the foaming pressure of the blowing agent present which generates a low level dwell pressure. Process control may be insufficient due to underdosing or overdosing of the melt combined with inaccuracies of the plunger stop.

A better process control is accomplished by applying a controlled dwell pressure with the help of the driving plunger. In order to prevent foam collapse, these pressures have to be one or two orders of magnitude smaller than the pressures applied during standard die casting processes.

There are essential differences between the processing of metals and polymers. The metal process is characterized by the fact that the blowing agent is added to the melt during the injection of the liquid metal into the steel mold (Fig. 2.6). Thus, the contact time between the liquid metal and the blowing agent until the mixture reaches the mold cavity is very short, about 20–50 ms. Accordingly, premature gas release from the blowing agent during form filling is minimized (see Sect. 2.5.2.2).

As mentioned before, the low viscosity of liquid metals has strong influence on the mold filling behavior, see Fig. 2.3. As a result, the form filling behavior during low pressure molding of metals (Fig. 2.7) is quite different from that of polymers (Fig. 1.8). The LP-IFM process is thus characterized as follows:

- Form filling and foam evolution are successive.
- The compact skin forms during form filling due to the chilling effect of the cold[16] mold wall. A good surface quality demands the mold wall to be completely wet by the molten metal during mold filling. LP-IFM is thus best suited for compact castings with short flow paths.
- Foam formation fills the core of the castings. This is different from polymers where extremities of the mold are filled by foam expansion.

The success of the integral foam molding process is thus strongly determined by the form filling behavior of the molten metal. High local filling velocities (10^1–10^2 m/s) combined with the chilling effect of the mold wall lead to rather good surface qualities of the castings which are – if wetting is complete – comparable to that of compact die castings.

An example for a magnesium LP-IFM demonstration part is shown in Fig. 2.8. The thickness of the dense skin is about 1–2 mm. The cellular core is quite homogeneous with a more or less constant mean pore diameter. There is no indication for a minimum relative density in the center of the casting.

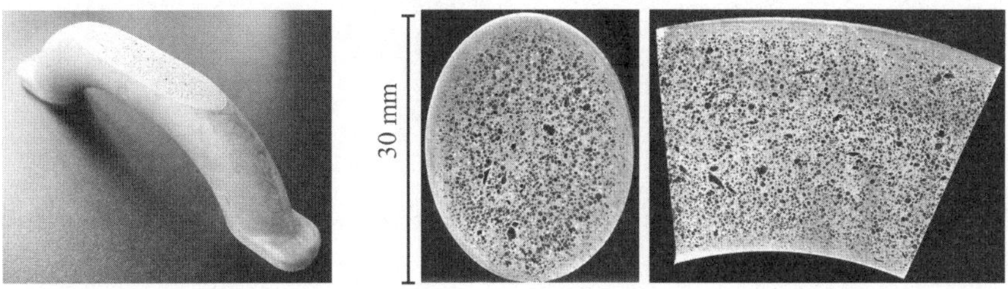

Figure 2.8: Foamed magnesium handle as an example for LP-IFM. Weight reduction: 30% (LP-IFM, thixomolding, magnesium alloy AZ91, blowing agent: MgH_2).

Handles with different mean relative density are depicted in Fig. 2.9. The density variation is realized by a different level of underfilling of the mold. Mean relative densities of about 0.6 may be realized while maintaining a high surface quality. At lower relative densities, cold welds, shrink holes and eventually sink marks appear.

The second example for LP-IFM shows that also large, plain geometries may be realized, see Fig. 2.10. The aluminum plate shown in Fig. 2.10 was produced by die casting with a dwell pressure maintained at 5 bar during solidification. There is no

[16]The mold temperature is between 200 and 300°C which is cold compared to the molten metal with a temperature of about 600–700°C.

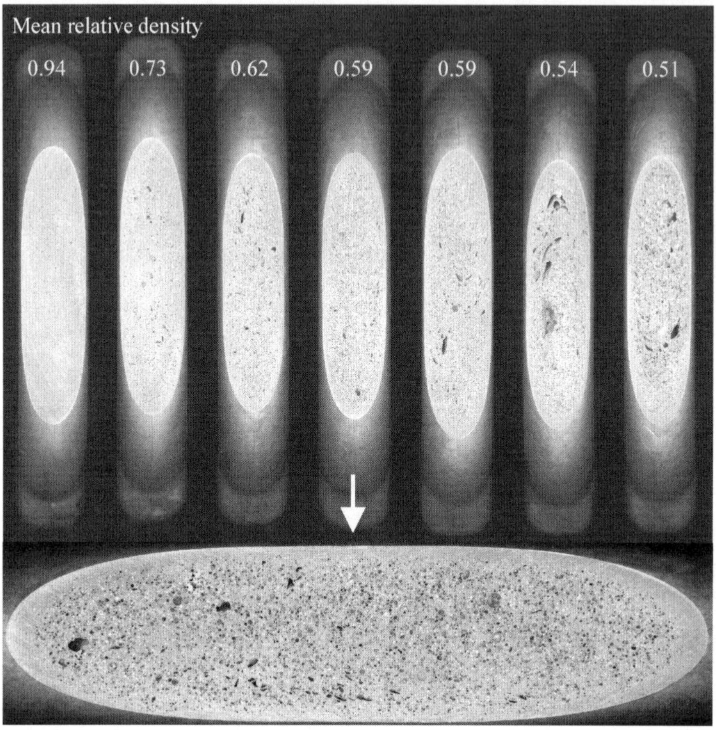

Figure 2.9: Foamed magnesium handle from Fig. 2.8 with decreasing mean relative density from left to right.

principal difference in the processing of aluminum and magnesium. Generally, the time available for foam formation is shorter for magnesium than for aluminum.

A clear benefit of LP-IFM is its intrinsic simplicity. The necessary modifications of the permanent steel mold for the admixing of the blowing agent are insignificant. The casting process is analogous to standard die casting with respect to melt temperature, mold temperature and casting velocity. The only difference is that the mold is not completely filled during molding. To apply a controlled dwell pressure of about 5–50 bar is not absolutely necessary but strongly increases the robustness of the whole process.[17] Standard die casting machines are not able to apply a con-

[17] The casting pressure generated by the dwell pressure is balanced by the foaming pressure of the blowing agent. The porosity is determined by the amount of blowing agent and the applied dwell pressure. Both, the amount of gas and the pressure, are linked via the ideal gas law. For details see Chap. 3.

Figure 2.10: Staged aluminum plate (10 mm, 12 mm, 14 mm) and cross-sections thereof. The porosity is about 0.4 (LP-IFM, die casting, aluminum alloy AlSi9Cu3, blowing agent: MgH_2, dwell pressure: 5 bar, plate dimension: 180 mm × 180 mm).

trolled dwell pressure of about 5–50 bar. Nevertheless, the necessary modifications of the die casting machine are also not extensive.[18]

Shortcomings of LP-IFM are the roughness and visible defects at the surface if the relative density falls below a critical value. This critical value is a function of the geometry of the casting and increases with decreasing wall thickness and increasing complexity of the molded part. As a rough indication, sound castings with a porosity of 0.2–0.4 can be realized. The danger to obtain visible sink marks or bad surfaces increases with higher porosities. These deficiencies are the result of an incomplete wetting of the mold wall during mold filling combined with an insufficient foaming pressure produced by the blowing agent.

Generally, LP-IFM is best suited for rather compact castings with short flow paths and moderate thickness variations. If large thickness variations, very long flow paths or partially foamed parts have to be realized, another molding strategy is better suited, see Sect. 2.4.

[18]We use the FRECH die casting machine DAK 450-54 modified by FRECH in order to control a low dwell pressure of about 5 bar.

2.4 High Pressure Integral Foam Molding

During low pressure IFM [153,235] the mold cavity is not completely filled and only a low dwell pressure is applied. Thin walled castings with long flow paths are thus difficult to realize. The philosophy of high pressure integral foam molding (HP-IFM) is to fill the mold completely and to apply a standard dwell pressure. After a short time delay, the plunger is fastened and foaming is initialized by local expansion of the mold.

The process principle of integral foam molding with expanding molds is depicted in Fig. 2.11. The basic method is analogous to the polymer process described in Sect. 1.2. The liquid metal is injected into a steel mold under standard die casting conditions. The mold is completely filled and a high dwell pressure of some 100 bar is applied. After a time delay t_{delay} of 0–100 ms, the plunger is stopped and the mold is locally expanded within about 250 ms thereby inducing foam evolution.

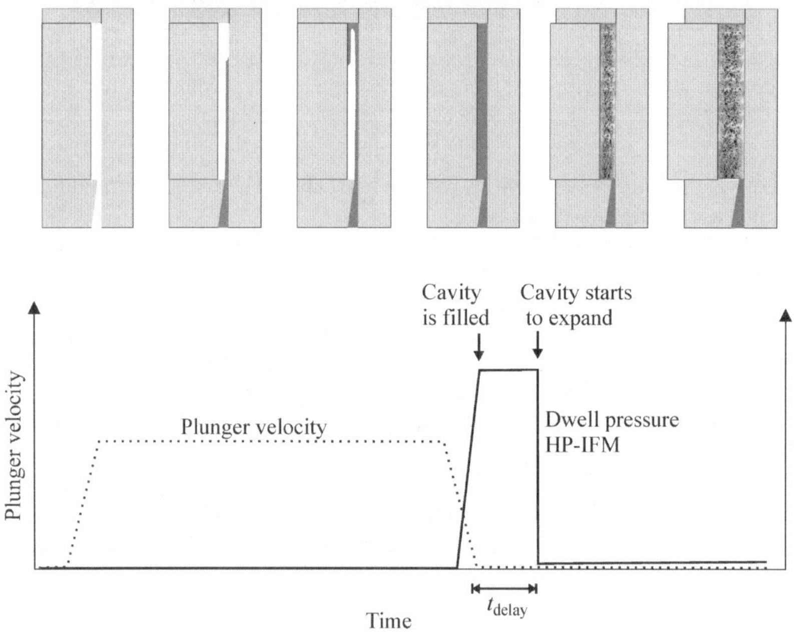

Figure 2.11: High pressure integral foam molding of metals (schematic). *Top*: Filling and expansion of the cavity. Bottom: Plunger velocity and dwell pressure.

The main differences between the metal and the polymer process are the involved times and temperatures. The solidification velocity of metals is much higher than that of polymers. After the mold is completely filled there is not much time left until expansion has to take place. The time delay, t_{delay}, for metals has to be a small fraction of a second whereas there is much more time available during polymer processing.[19]

Generally, some machine modifications are necessary to realize HP-IFM with standard die casting machines. On the one hand, a plunger stop function is required to prevent the plunger from compressing the foam during mold expansion. On the other hand, a very fast coupling between the hydraulic cylinder and the plunger is crucial, see Fig. 2.12.[20]

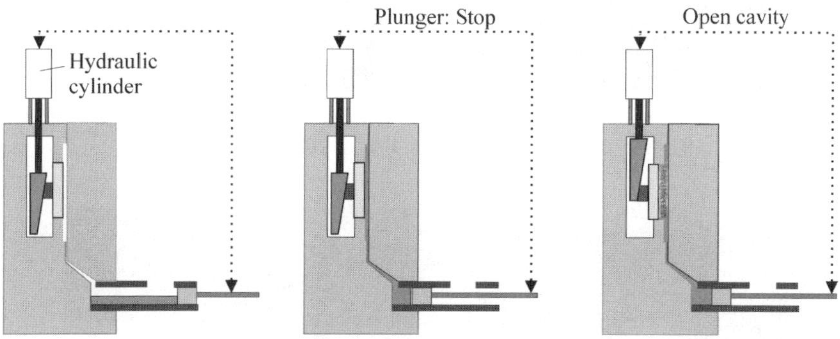

Figure 2.12: Schematic sketch of the mold design and machine coupling for HP-IFM.

A HP-IFM mold is depicted in Fig. 2.13. The plate with dimension 175 mm × 175 mm shows a movable core with dimension 150 mm × 150 mm. Expansion is possible starting from 2 mm up to 18 mm.

A lateral view of a HP-IFM part cast with the mold shown in Fig. 2.13 is depicted in Fig. 2.14. The solidified surface sticks to the mold wall and follows obviously its movement. As a consequence, a gap – for this experiment 4 mm – opens which appears as a bright strip on the component. The gap is refilled with foaming melt during mold expansion. The new surface skin is fairly rough and comparable with the surface of metal foams produced by e. g. foaming of powder compacts [2, 17].

There are some specific advantages of HP-IFM compared to LP-IFM:

- Intricate elements and thin-walled structures with long flow paths can be realized.

[19]Typical delay and expansion times for polymers are 2 s and 6 s, respectively [183].

[20]These modifications were successfully performed by FRECH for the FRECH DAK 450-54. The stop of the plunger is synchronized with the controlled expansion of the mold.

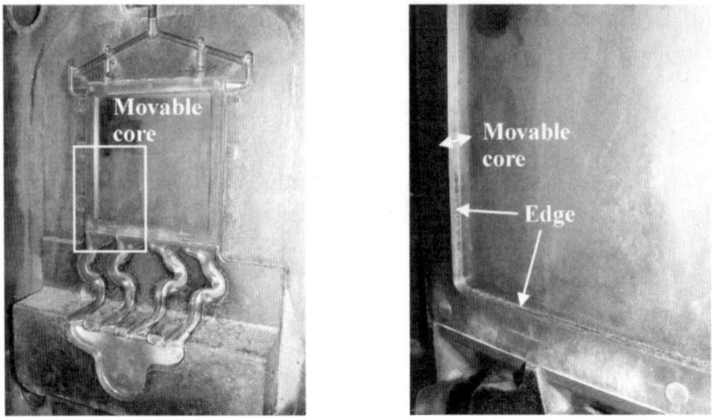

Figure 2.13: HP-IFM mold. Plate with movable core.

Figure 2.14: HP-IFM part. (**a**) General view, (**b**) lateral view: without expansion (*left*) and with 4 mm expansion (*right*), (**c**) cross section. (HP-IFM, die casting, aluminum alloy AlSi9Cu3, blowing agent: MgH_2)

- Locally foamed castings are possible since mold expansion can be local.
- The thickness of the skin can be varied by varying the time delay t_{delay}.
- Generally, higher porosities can be realized compared to LP-IFM.

Nevertheless, there are also drawbacks:

- Mold design is more complex. Due to the movable elements, there are limitation with respect to the part geometry.
- There are visible surface marks where mold expansion takes place.

2.5 Base Materials

The success of IFM is not only dependent on the processing strategy but as well a function of the utilized materials. Appropriate matrix alloys and blowing agents have to be selected.

2.5.1 Matrix Alloys

Die casting light metal alloys are optimized to produce sound castings and not integral foams. Special polymers without or with additives are used to produce polymer integral foam. What kind of metal alloys are recommended for IFM?

Generally, pure melts do not foam. The first step characteristic for all metal foaming technologies is thus to condition the base metal alloy e. g. by adding particles or by oxidation.[21] No matter what kind of strategy is pursued, the underlying mechanism which makes metal melts foamable is always the same. There are particles present which act as obstacles in cell walls and prevent or slow down cell wall rupture.[22]

Fortunately, conditioning of the melt or the use of particle reinforced melts has shown to be not necessary for IFM. Essential for the foaming result is the solidification behavior of the underlying alloy. The solidification morphology is not only dependent on the alloy but also on the solidification velocity and the involved temperature gradients. The location of nucleation and the way how growth takes place determines the solidification type. Generally, exogenous and endogenous solidification types are distinguished, see Fig. 2.15 [105, 130].[23]

During exogenous solidification, nucleation starts at the mold wall and nuclei growth is in direction of the thermal center of the casting. In that case, smooth, rough and sponge-like solidification are distinguished. Endogenous solidification designates the situation when nucleation starts in the melt and a mixture of liquid and solid phase forms. In this case, mushy and shell-forming solidification are distinguished.

Standard die casting alloys are normally near-eutectic and show an endogenous solidification type. The grain size depends on the thickness of the casting and is about 10µm [130]. Thus, foam formation takes place in a liquid containing a huge number of tiny solid phase nuclei. The solid phase particles play the same role as the particles which are deliberately added during other metal foam production processes.

---

21 For details, see Sect. 4.3.

22 This stabilization mechanism is discussed in detail in Chap. 6.

23 This classification is analogous to columnar and equiaxed [164]. However, endogenous and exogenous describe the underlying situation much better.

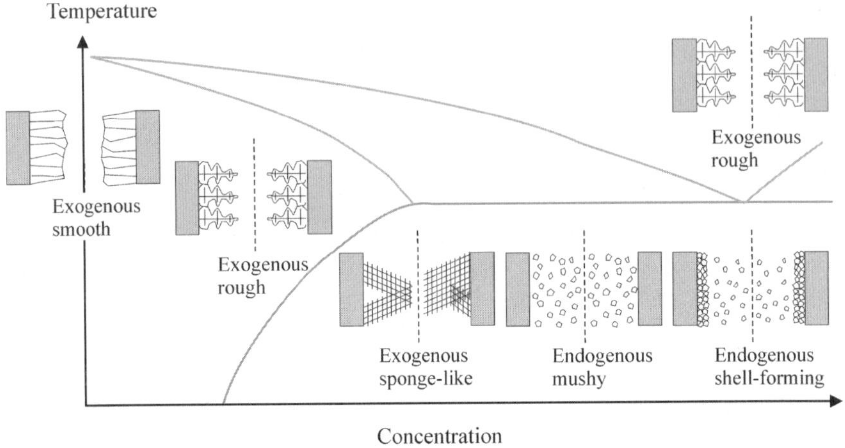

Figure 2.15: Solidification categories of casting alloys in the eutectic system. (following Engler [65, 66])

Summarizing, during IFM standard die casting alloys, e. g. the aluminum alloy AlSi9Cu3 or the magnesium alloy AZ91, can be used.[24] We do not have to add any additions to the melt. This is a huge advantage with respect to costs, foam properties and recycling.

Aluminum as well as magnesium alloys can be processed to integral foam parts. Nevertheless, integral foam molding is a race against time. Foam formation is no longer possible if solidification proceeds faster than the release of the blowing gas. The heat content of magnesium alloys is generally much lower than that of aluminum alloys, about a factor of two.[25] Consequently, the processing of magnesium is more difficult than that of aluminum due to the much shorter time interval available for foam formation. This is true for both molding processes described in this chapter. However, the resulting structures and properties appear to be not dependent on the light metal used if the aspect of the different solidification velocity is taken into account. This is the reason why we usually do not explicitly distinguish between aluminum and magnesium in the following. Generally, results for aluminum are also valid for magnesium and vice versa.

[24] AlSi9Cu3: Si: 8.0–11.0%, Cu 2.0–3.0%, Mn: 0.1–0.5%, Mg: 0.1–0.5%, Al: balance.
AZ91: Al: 8.0–9.5%, Zn: 0.3–1.0%, Mn: 0.1–0.3%, Mg: balance.

[25] The heat of solidification is about 550 kJ/kg and 380 kJ/kg for AlSi9Cu3 and AZ91, respectively [73].

2.5.2 Blowing Agents

Indispensable for the success of the integral foam molding process is the availability of an adequate blowing agent which should show the following properties:

- **A high equilibrium dissociation pressure.** The foaming pressure rises with the equilibrium pressure of the blowing agent.
- **A high storage capacity for gas.** The higher the storage capacity the lower is the amount of blowing agent which has to be added.
- **A fast decomposition kinetics.** The decomposition kinetics determines the efficiency of the blowing agent.
- **No harmful influence on the casting alloy.** The addition of the blowing agent changes the composition of the alloy. Elements which might lead to a degradation of the alloy or which make recycling difficult have to be avoided.

Many different blowing agents have been tested as blowing agents for metals: TiH_2, SrH_2, MgH_2, $CaCO_3$ [51,79,81]. Up to now, the best results were achieved with metal hydrides, especially TiH_2, whose decomposition temperature is close to the melting point of aluminum. Even so, TiH_2 is not suited for the integral foam molding process due to its too slow decomposition kinetics at casting temperatures [122]. For the purpose of integral foam molding, MgH_2 has been shown to be much more suitable.

In the following, thermodynamics and kinetics of MgH_2 decomposition is discussed in detail.[26] Thermodynamics provides information about the obtainable foaming pressure whereas kinetics determines the relevant time scale for the decomposition of the blowing gas.

2.5.2.1 Thermodynamics

Many metals (Me) react with hydrogen to form metal hydrides according to

$$\mathrm{Me} + \frac{\mathrm{x}}{2}\mathrm{H_2} \longleftrightarrow \mathrm{MeH_x}. \tag{2.2}$$

Metals are able to absorb large amounts of hydrogen with a very high volumetric hydrogen density, even higher than that of liquid hydrogen. Especially light metal hydrides, e. g. MgH_2, show a high storage capacity by weight, see Table 2.1.

Generally, the reaction is exothermic and reversible, i. e. hydrogen is desorbed again by heat treatment. The hydrogen concentration in the α phase, which denotes the metal hydrogen solid solution, is pressure dependent and can be described by

[26] MgH_2 from Goldschmidt AG, Essen (Tego Magnan [90]).

Table 2.1: Properties of various hydrides [204].

	Density (g/cm^3)	Wt % H	Volumetric H density (atoms $H/cm^3 \cdot 10^{22}$)
H_2, liquid (20 K)	0.071	100	4.2
MgH_2	1.40	7.6	6.7
TiH_2	3.80	4.0	9.1

Sieverts' law.[27] As the hydrogen pressure p_{H_2} increases, saturation of the matrix occurs and the β phase, which is the metal hydride MeH_x, starts to form:

$$MeH_y + \frac{1}{2}(x - y)H_2 \longleftrightarrow MeH_x. \tag{2.3}$$

The conversion to the hydride phase β takes place at constant pressure in accordance to Gibbs' phase rule.[28] A schematic pcT-diagram and van t'Hoff plot is depicted in Fig. 2.16.

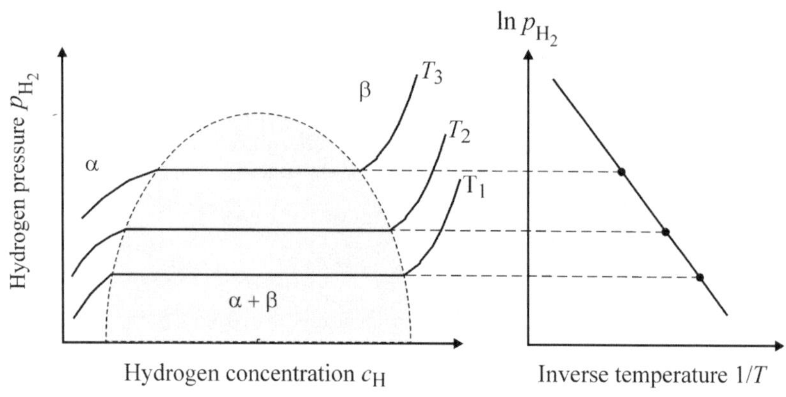

Figure 2.16: Schematic pcT-diagram and van t'Hoff plot. The solid solution phase and the hydride phase are denoted by α and β, respectively.

The temperature dependent plateau pressure is the equilibrium dissociation pressure and a measure for the stability of the hydride. The equilibrium dissociation

[27] See p. 80.

[28] Gibbs' phase rule, $P + F = B + 2$, correlates the number of phases P, the number of free state variables F and the number of different components B.

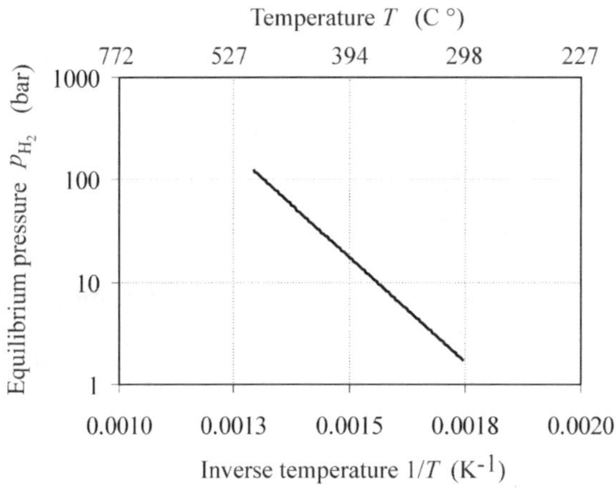

Figure 2.17: Van t'Hoff plot for MgH_2 [90, 207]. The equilibrium dissociation pressure reaches pressures of several hundred bar at temperatures above 500°C.

pressure can be described by the well-known **van t'Hoff equation**

$$\ln p = \frac{\Delta H}{RT} - \frac{\Delta S}{R} \tag{2.4}$$

where ΔH and ΔS represent the enthalpy and entropy change, respectively. Figure 2.17 shows the van t'Hoff plot for MgH_2. Pressures of several hundred bar are reached for temperatures above 500°C.[29] These quite high pressures are very important since they are a measure for the foaming pressure generated by the blowing agent during IFM.

2.5.2.2 Kinetics of Magnesium Hydride Decomposition

The blowing agent decomposition takes place under extreme conditions during IFM. The MgH_2[30] is mixed within 20–50 ms with the about 600°C hot molten metal. That is, the blowing agent is heated up very quickly. After mixing there is only a small time interval available until solidification puts an end to the foaming process. Near the mold walls, solidification occurs more or less instantaneously. In the interior of the casting, it takes some hundred milliseconds up to a few seconds until solidification is terminated. During solidification, the temperature varies only moderately in the

[29] These pressures are similar to the pressures typical for the MuCell™ technology, see Sect. 1.2.3.

[30] The particles are flake-like with mean diameters of 100 μm and thicknesses of about 10–20 μm, see p. 50.

two-phase region. In a simplified picture, the physical situation may be described by an instantaneous heating of the blowing agent particles to the melt temperature and a subsequent holding at this temperature for a period of time.

Metal hydrides were originally designed for compact high density hydrogen storage. Consequently, the temperature conditions for which the decomposition of metal hydrides is usually investigated are far away from those which are relevant for IFM [204]. Experimental information about the decomposition behavior of MgH_2 under conditions typical for IFM is non-existent in literature.

In the following, a method is described to extract information about the high temperature behavior of MgH_2 decomposition from decomposition experiments at constant heating rate. Experimentally, the decomposition of MgH_2 is characterized by thermo-gravimetry. Obviously, a heating rate in accordance to IFM is far beyond the experimental possibilities. However, the behavior at these extreme conditions may be extrapolated from measurements at heating rates which are experimentally accessible if the experimental information is analyzed on the basis of an adequate decomposition theory. During this procedure, all material parameters are determined and the underlying theory can subsequently be applied on the experimental situation of interest.

Theoretical basis for the description of the decomposition is a Johnson-Mehl-Avrami (JMA) type nucleation and growth rate equation which is based on the assumption that the dehydriding process is a nucleation and growth phenomenon. Nucleation and growth of the α phase from the β phase is assumed to represent the governing process during MgH_2 decomposition. For special cases of nucleation and growth it is possible to derive the well-known analytical description of transformation kinetics according to Johnson, Mehl and Avrami, the **Johnson-Mehl-Avrami equation** [6,129,204],

$$f(t) = \exp\left[-(kt)^n\right], \tag{2.5}$$

where f is the phase fraction of the decaying phase and t is the elapsed time. The Avrami exponent n and the effective kinetic parameter k are constants which have to be experimentally determined. For the temperature dependence of k an Arrhenius expression is assumed,

$$k = k_0 \cdot \exp\left(-\frac{E_\mathrm{A}}{RT}\right), \tag{2.6}$$

where k_0 is a constant and E_A the apparent activation energy.

In order to apply JMA-theory, Eq. (2.5), to the experimental situation under consideration, three material constants have to be determined: E_A, k_0 and n. This is done by thermo-gravimetry where the samples are heated with constant heating rates much lower than during IFM. Thus, the JMA theory, which describes the decomposition at constant temperature, has to be generalized in order to be applicable on thermo-gravimetry experiments with constant heating rate, see Appendix A.1. With the material parameters determined in Appendix A.1 we are able to apply the

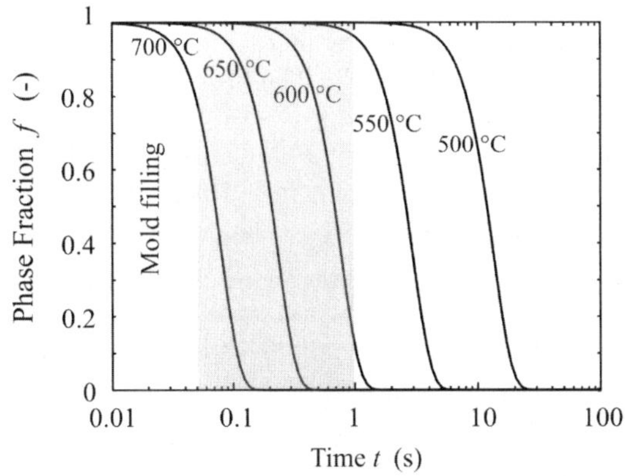

Figure 2.18: Decomposition of MgH_2 at different constant temperatures based on JMA theory, Eq. (2.5) and Eq. (2.6), using the parameters from Table A.1. The shaded region indicates the relevant time scale for IFM.

JMA theory. Figure 2.18 shows the calculated temporal decomposition of MgH_2 at different temperatures. The shaded region in Fig. 2.18 indicates the relevant time interval during die casting or injection molding. From this theoretical result and the fact that typical casting temperatures are between 600 and 700°C it follows:

- The temperature dependence of the decomposition is very pronounced.
- The duration and the temperatures of the casting process are coherent with the decomposition dynamics of the MgH_2.
- Premature gas evolution during mold filling is – if even present at all – weak.

In summary, the high dissociation pressure combined with the very fast dissociation dynamics at the relevant temperatures makes MgH_2 an ideal choice as a blowing agent for light metal IFM.

2.6 Synopsis

Integral foam molding of light metals is in fact possible. In the following, the most important insights are summarized:

- The low viscosity of metal melts has emerged as an advantage. It allows turbulent mixing in the permanent steel mold during the shot. In addition, the resultant surface quality is good since metals are sprayed.

- Two different integral foam molding processes have been developed:
  (i) LP-IFM: During low pressure integral foam molding, a permanent steel mold is partly filled with melt. The pressure of the evolving blowing gas fills the residual hollow space with foam.
  (ii) HP-IFM: An expandable steel mold is completely filled with metal melt with standard die casting parameters. Foam formation is locally provoked by expansion of the mold after a delay time of about $0 - 100$ ms.

  Both methods show specific advantages and disadvantages.

- Standard die casting alloys may be used. Additives or a pretreatment of the melt are not necessary. Recycling is no issue.

- Magnesium hydride is a suitable blowing agent for aluminum and magnesium IFM. It shows ideal decomposition characteristics in the relevant time and temperature interval. In addition, the equilibrium dissociation pressure reaches very high values of over 100 bar and its hydrogen storage capacity is enormous.

- Processing of magnesium is more difficult than that of aluminum due to the lower heat content. Especially for HP-IFM, the time delay should be very short and the expansion time should be as short as possible.

2.6.1 Design Guidelines

Generally, it is not reasonable to substitute standard die casting parts simply by foamed parts. In order to produce sound castings and to exploit the advantages of integral foam, a complete new design of the component is generally necessary. Some important guidelines are as follows:

- LP-IFM is best suited for compact, thick-walled parts with a low degree of complexity.
  - Flow paths should be as short as possible.
  - The minimum foamable wall thickness is about 4 mm. Thinner local structure elements such as ribs or connection elements may also be realized if the flow paths are short.
  - Thickness variations of the casting should be moderate and square-edged design should be avoided.

- HP-IFM is best suited for flat parts and parts with a local foam core.
  - There are no restrictions with respect to wall thicknesses, connection elements, flows paths, etc.

- There are restrictions with respect to geometry since one or more core pullers have to be placed in the steel mold.
- HP-IFM castings show visible surface marks which have to be taken into consideration.

2.6.2 Costs

The production costs of integral foam parts are comparable to standard die castings. The costs for the blowing agent[31] are nearly compensated by material savings due to the density reduction. The cycle times are similar to standard die castings. There are only minor modifications necessary for blowing agent feeding. For HP-IFM, additional costs originate from the necessity of core pullers. Minor modifications of the die casting machine, control of a low dwell pressure for LP-IFM and plunger stop and very fast expansion of the cavity for HP-IFM, are necessary. In summary: Integral foam parts are not expensive!

[31] 1 kg $MgH_2 \approx 180$ €

3 Structure and Properties

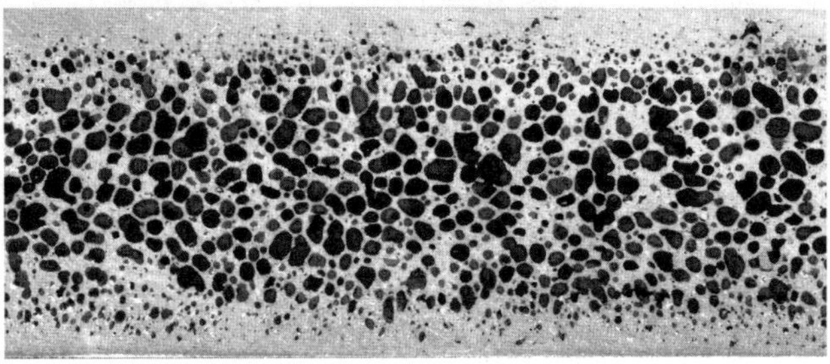

Integral foams are not materials but structures!

The main characteristics of integral foam castings are a compact skin with a gradual transition to a foamed core. The skin gives the casting its form and strength while the core can be used to optimize the weight-specific bending stiffness, the energy absorption capacity or the damping properties.

3.1 Density Profiles

The outstanding feature of integral foam manifests itself in its gradual transition from a compact outer skin to a foamed core. The material distribution not only determines the mechanical properties of the foamed part (see Sect. 3.3) but also reveals the basic processes and physical mechanisms during foam evolution.

3.1.1 Experimental Observations

Integral foams may must not be treated as homogeneous materials but have to be regarded as structures, see Fig. 3.1.[1]

[1]Even so, material properties like the bending stiffness or the damping behavior are dominated by the relative density, see Sect. 3.3.

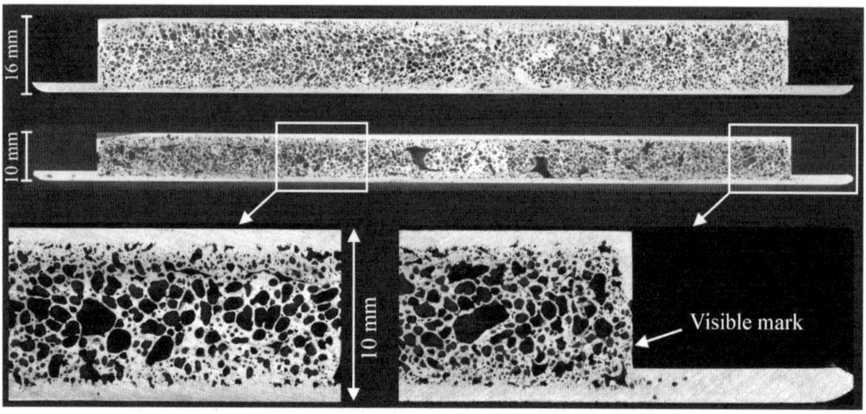

Figure 3.1: Cross-sections of HP-IFM castings. *Top*: Mold expansion from 8 to 16 mm. *Bottom*: Mold expansion from 6 to 10 mm. (die casting, aluminum alloy: AlSi9Cu3, blowing agent: MgH_2, component measures: $175 \times 175 \times 10\ mm^3$ and $175 \times 175 \times 16\ mm^3$)

A suitable approach to characterize integral foam is to determine density profiles.[2] Figure 3.2 shows two typical density profiles. The thickness of the compact skin is about one mm. There is a continuous but quite sharp transition to the foamed core. The curve progression is symmetric and shows a pronounced plateau with minimum density midway.

The skin is in fact dense. In contrast to polymer integral foam molding, see Sect. 1.2, there are no indications for surface bubbles compressed by the dwell pressure. Premature gas release from the blowing agent and foam evolution during mold filling do not take place. This observation is in accordance with the decomposition kinetics of MgH_2 discussed in Sect. 2.5.2. Thus, mold filling and foam evolution are sequential during IFM, which is essential in view of a sound surface skin.

There is a minimum relative density which can be realized with IFM. Generally, HP-IFM allows lower relative densities than LP-IFM. Trying to achieve very low relative densities during LP-IFM leads to visible marks at the surface of the casting as a result of an incomplete wetting of the mold wall. The non-wet areas have to be filled by the foaming melt. Even if the foaming pressure is sufficient to fill the mold cavity, the surface quality of these areas is strongly reduced, see Fig. 2.4.

[2] Density profiles are determined by micro computed tomography [193, 203]. We use a μCT40 from Scanco Medical and a voxel size of $(10\ \mu m)^3$.

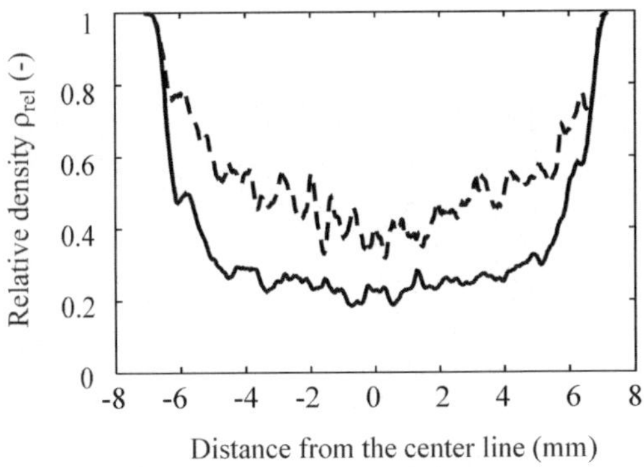

Figure 3.2: Density profiles for two HP-IFM aluminum castings. *Dashed line*: Mold expansion from 8 to 15.4 mm, *Full line*: Mold expansion from 6 to 15.4 mm. (μCT measurement)

During HP-IFM the mold is first completely filled before foam formation is initiated by mold expansion. Wetting of the mold walls is thus not an issue. Relative core densities of about 0.2 and lower can be realized. Limitations appear when the foaming material is no more able to close the evolving gap during mold expansion.

3.1.1.1 Influence of the Wall Thickness

Generally, the attainable relative density is a function of the thickness of the cast part. Higher thicknesses allow higher porosities. There is a lower limit of about 3–4 mm which can be reasonably foamed due to the interplay between decomposition kinetics and solidification.[3] The influence of the thickness of the casting on the density profile is depicted for LP-IFM in Fig. 3.3. With increasing thickness, the appearance of the density profile changes from Gaussian to a pronounced plateau region. Despite that, the thickness of the compact skin, which is a manifestation of the time delay of the decomposition of the blowing agent, seems not to be very much influenced by the thickness of the casting and is always about 1–1.5 mm.

3.1.1.2 Influence of the Relative Density

The influence of the relative density on the density profile is depicted in Fig. 3.4 for 6 and 10 mm wall thickness. The density profile changes with decreasing relative

[3]This limit may decrease if blowing agents are available with faster decomposition dynamics compared to MgH_2. In addition, higher mold temperatures or thermal coatings may help to lower these limits.

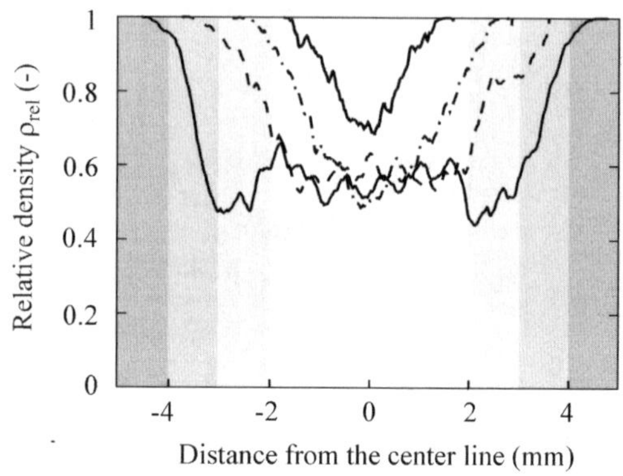

Figure 3.3: Density profiles of LP-IFM castings with 4, 6, 8, and 10 mm wall thickness. (μCT measurement, thixomolding, magnesium alloy AZ91)

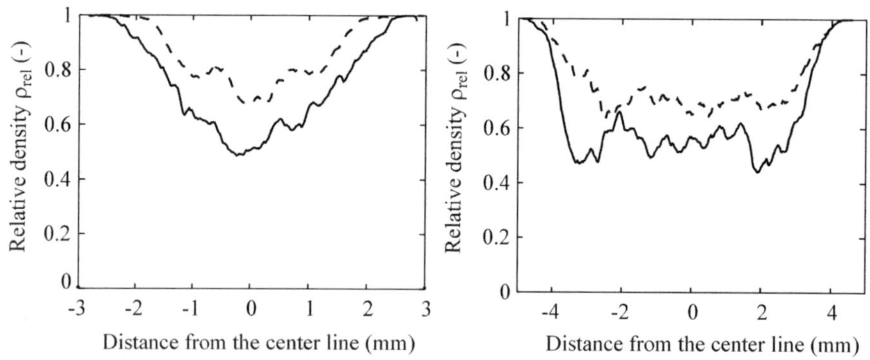

Figure 3.4: Density profiles of LP-IFM magnesium castings with 6 mm (*left*) and 10 mm (*right*) thickness for different relative densities. (μCT measurement, thixomolding, magnesium alloy AZ91)

density but the functional dependence of the profiles at constant thickness appears similar.

3.1.1.3 Correlation Between Relative Density and Casting Pressure

A further important observation is the dependence of the relative density of integral foam castings on the applied casting pressure p_{cast}, see Fig. 3.5.

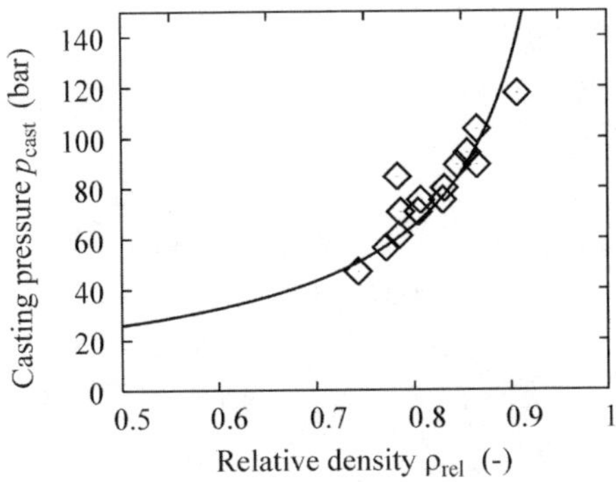

Figure 3.5: Correlation between casting pressure p_{cast} and resultant relative density ρ_{rel}. The data points belong to a ring-shaped aluminum part with diameter of about 100 mm and wall thickness between 3 and 8 mm. *Full line*: Eq. (3.1). (LP-IFM, die casting, aluminum alloy AlSi9Cu3)

The applied casting pressure balances the foaming pressure generated by the released blowing gas which is governed by the ideal gas law. If the amount of decomposed blowing agent is constant, the relative density is intimately connected with the the casting pressure p_{cast}:

$$p_{cast} \propto p_{dwell} \propto \frac{1}{V_G} \propto \frac{1}{1 - \rho_{rel}}, \tag{3.1}$$

where V_G denotes the total pore volume of the casting. Equation (3.1) provides the possibility to controll the molding process and especially the resultant porosity with the help of the dwell pressure.

3.1.1.4 Residual Blowing Agent Particles

Further evidence for the importance of the decomposition dynamics is the distribution of residual blowing agent particles in the casting, see Fig. 3.6.

In electron microscopy pictures the flake-like MgH_2 particles appear dark gray compared to the aluminum matrix. There is a strong correlation between the presence of the MgH_2 particles and the porosity. With increasing distance to the surface, the number and the area of these particles drops drastically due to decomposition and dissolution in the aluminum matrix.

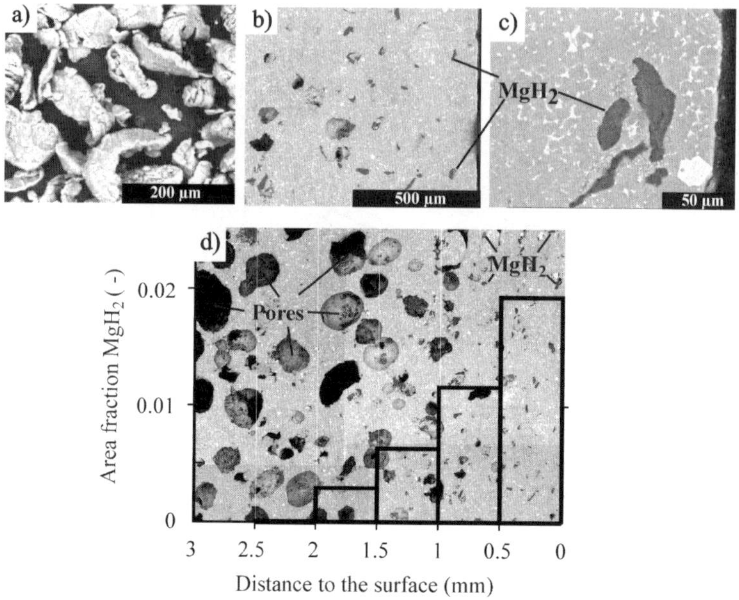

Figure 3.6: Residual MgH_2 particles in a 6 mm LP IFM aluminum casting. (**a**) MgH_2 particles as received, (**b**) MgH_2 particles embedded in the AlSi9Cu3 matrix, (**c**) MgH_2 particles near the surface, (**d**) Area fraction MgH_2 as a function of the distance to the surface.

3.1.2 Theoretical Considerations

The aim of this section is to develop a deeper theoretical understanding of the integral foam molding process based on simple but sufficient physical models.

3.1.2.1 Density Profiles

Integral foam molding is characterized by a competition between solidification and foam formation. The main feature of the molded structures – the density profile – is already reproduced by a thermal model combined with the decomposition dynamics of the blowing agent (Sect. 2.5.2.2). The underlying assumptions for the theoretical model[4] are as follows:

[4]This model is the simplest model that can be imagined. However, although simple it shows to be sufficient to understand the main principles.

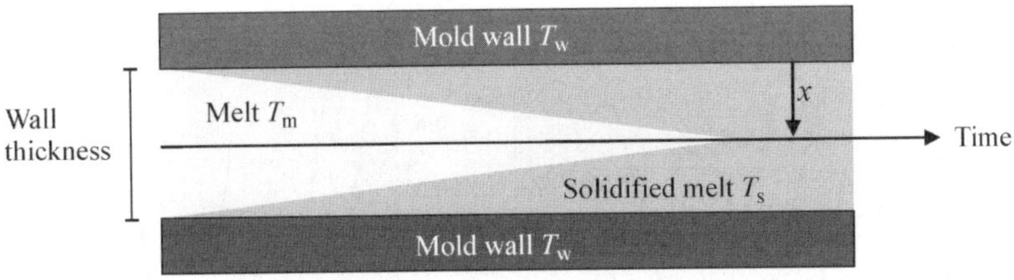

Figure 3.7: Thermal model for solidification.

- The system can be characterized by three constant temperatures, see Fig. 3.7: T_w: Mold wall temperature, T_m: Temperature of the melt, T_s: Temperature of the solidified melt.
- Solidification starts at the mold wall and ends at the center line. A two-phase region is not taken into account.
- The solidification rate $\dot{x}$ is determined by the heat transfer to the mold wall:

$$\dot{x} = \frac{(T_\mathrm{s} - T_\mathrm{w})\,\alpha_\mathrm{h}}{\rho\,Q_\mathrm{s}}, \tag{3.2}$$

  with the heat transfer coefficient α_h[5] and the heat of fusion Q_s.
- Super heating of the melt is ignored.
- Decomposition of the blowing agent at distance x from the mold wall is at constant temperature T_m and takes place as long as the solidification front has not reached x. The state of the blowing agent MgH_2 after solidification is characterized by the residual phase fraction $\tilde{f}(x)$:

$$\tilde{f}(x) \;:=\; f(x/\dot{x}) \;\overset{\text{Eq.(2.5)}}{=}\; e^{-(kx/\dot{x})^n} \qquad \text{for} \quad x \leq \frac{d}{2}, \tag{3.3}$$

  where d denotes the wall thickness.
- The resultant relative density profile $\rho_\mathrm{rel}(x)$ can be expressed as a function of the residual phase fraction $\tilde{f}(x)$, the casting pressure p_cast and the mass fraction of blowing agent particles ϵ_m.[6]

[5] Over the years, several researchers have attempted to determine heat transfer coefficients in permanent molds and have obtained some widely divergent values ranging from 20.000 $\mathrm{W/m^2K}$ to less than 1.000 $\mathrm{W/m^2K}$ [112].

[6] The mass fraction ϵ_m depends on the amount of blowing agent particles added to the melt ($\rho_{\mathrm{MgH_2}} = 1.4\,\mathrm{g/cm^3}$).

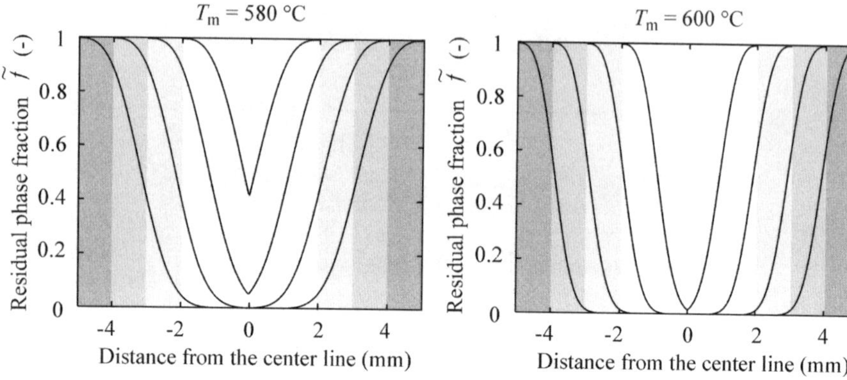

Figure 3.8: Decomposition of the blowing agent for $d = 4$, 6, 8 and 10 mm. Calculated residual phase fraction $\tilde{f}$ of MgH_2 (Eq. (3.3), Table A.1 and $\dot{x} = 1.6$ mm/s). *Left*: $T_m = 580\,°C$. *Right*: $T_m = 600\,°C$.

The gas volume V_G is proportional to the fluid volume V_F, ϵ_m, $(1 - \tilde{f}(x))$ and $1/p_{cast}$:

$$V_G = c\,\frac{V_F\,\epsilon_m\,(1-\tilde{f}(x))}{p_{cast}} \quad \Leftrightarrow \quad \frac{V_G}{V_F} \overset{\text{Eq.(1.3)}}{=} \frac{1}{\rho_{rel}} - 1 = c\,\frac{\epsilon_m\,(1-\tilde{f}(x))}{p_{cast}}, \tag{3.4}$$

where c is a constant.

From Eq. (3.4) the **density profile** $\rho_{rel}(x)$ can be deduced:

$$\rho_{rel}(x) = \frac{1}{1 + c\,\epsilon_m \frac{(1-\tilde{f}(x))}{p_{cast}}} \overset{\text{Eq.(3.3)}}{=} \frac{1}{1+\tilde{c}\,(1-e^{-(kx/\dot{x})^n})} \qquad \text{for} \quad x \le \frac{d}{2}, \tag{3.5}$$

where $\tilde{c} = c\,\epsilon_m/p_{cast}$ is a constant.

The residual phase fraction $\tilde{f}(x)$ of MgH_2 for $d = 4$, 6, 8 and 10 mm is depicted for $T_m = 580°C$ and $T_m = 600°C$ in Fig. 3.8.[7] There is a very strong sensitivity to T_m. Figure 3.8 reveals two important facts. On the one hand, decomposition is rather weak near the mold wall which explains the observation of rather compact surface skins. On the other hand, the phase fraction decreases steeply and eventually reaches a more or less pronounced plateau in the kernel. The latter behavior explains the experimental observation of a marked plateau of the porosity in the kernel of thick-walled castings.

[7]Parameters (AZ91 [73]): $\alpha_h = 5000$ J/(m^2 K s), $Q_s = 380$ kJ/kg, $\rho = 1.6$ g/cm^3, $T_s - T_w = 230°C$.

Calculated and measured density profiles are compared in Fig. 3.9 in order to demonstrate the significance of the model. Although very simple, the theoretical

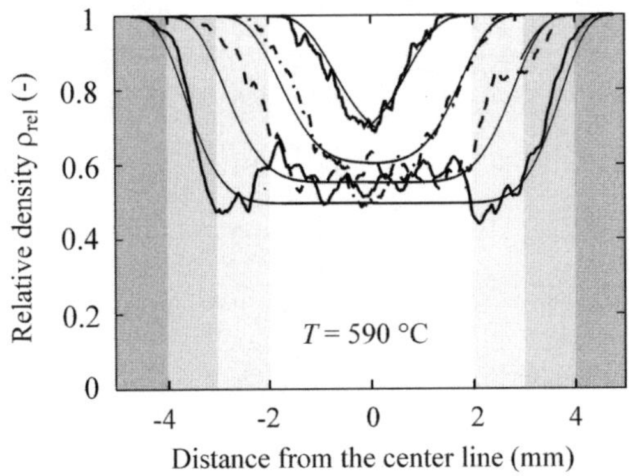

Figure 3.9: Density profiles from Fig. 3.3 compared with theoretical predictions, Eq. (3.5) (*thin, full lines* for $T = 590°\text{C}$ and $\dot{x} = 1.6$ m/s, 4 mm: $\tilde{c} = 0.5$, 6 mm: $\tilde{c} = 0.65$, 8 mm: $\tilde{c} = 0.8$, 10 mm: $\tilde{c} = 1.0$).

approach gives a qualitative reproduction of the experimental findings. Obviously, the decomposition dynamics of the blowing agent governs the evolution of the internal structure of the casting.

Also the experimentally observed conformity of the density profiles for different relative densities becomes clear on basis of Eq. (3.5). The relative density of the casting is controlled by p_{cast} and ϵ_{m}. The curve progression is determined by the decomposition dynamics of the blowing agent given by the residual phase fraction $\tilde{f}(x)$. Figure 3.10 shows experimental and theoretical density profiles for different wall thicknesses and relative densities.

3.1.2.2 Foaming Pressure

There are different reasons for the importance of the foaming pressure p. On the one hand, the foaming pressure might suppress the decomposition of the blowing agent when exceeding the equilibrium pressure. On the other hand, the foaming pressure is essential to understand the correlation between the applied dwell pressure and the resultant porosity of the casting.

We use a simple sandwich model, where only blowing agent particles present in the core of the casting contribute to the foaming pressure, to estimate p, see Appendix B.1. The foaming pressure given by Eq. (B.7) is depicted in Fig. 3.11.

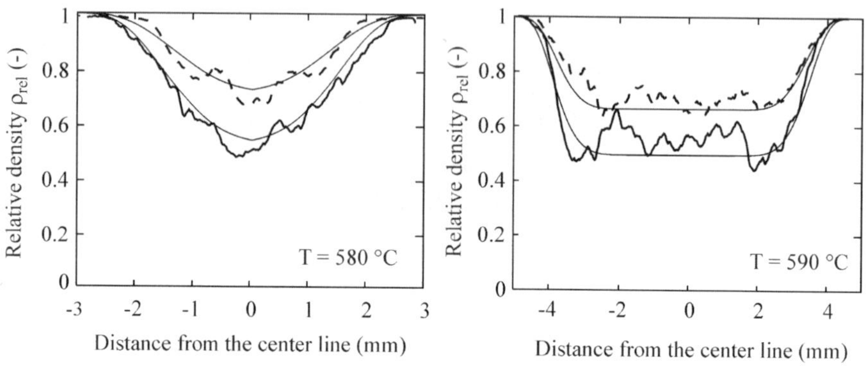

Figure 3.10: Density profiles from Fig. 3.4 compared with theoretical predictions, Eq. (3.5) (*thin, full lines*: $\dot{x}$ = 1.6 m/s. 6 mm (*left*): T = 580 °C, $\tilde{c}$ = 0.4 and 0.9. 10 mm (*right*): T = 590°C, $\tilde{c}$ = 0.5 and 1.0).

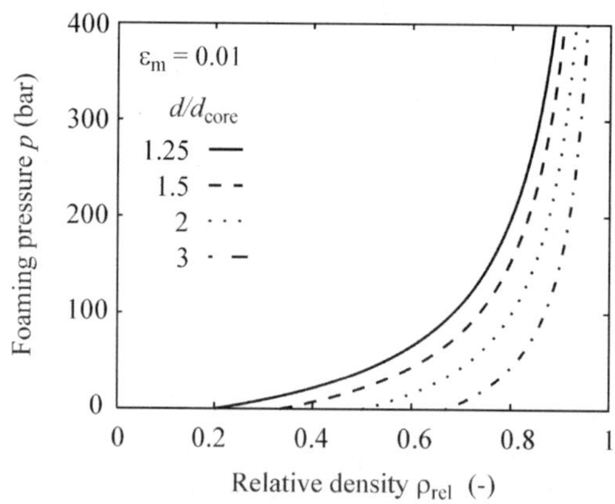

Figure 3.11: Calculated foaming pressure for aluminum, Eq. (B.7), as a function of the relative density of the casting for different d/d_{core} (d: thickness of the casting, d_{core}: thickness of the core) and $\epsilon_{\text{m}} = 0.01$.

Generally, the calculated pressures are at the same level as the applied casting pressures, see Fig. 3.5. Rather high foaming pressures of some 10^2 bar develop during IFM! These high pressures are the origin of the high surface quality of the castings. For HP-IFM, the pressure guarantees that the solidified surface follows the moving core. Although rather high, the estimated pressures for low relative densities are still below the equilibrium dissociation pressure of MgH_2 which is about 500 bar at 590°C, see Fig. 2.17. Generally, there seems to be not much danger to suppress decomposition of the blowing agent. Nevertheless, there is a clear possibility to suppress blowing agent decomposition for high ϵ_{m} or very high relative densities.[8] The calculated foaming pressure also reveals why thin castings (d/d_{core} is large) are difficult to realize, especially for low relative densities. The foaming pressure might not be sufficient to fill the mold before solidification puts an end to process.

3.2 Foam Structure Evolution

There are two reasons why the investigation of the foam structure evolution is important. Firstly, the structure has direct influence on the mechanical and physical properties. Secondly, and this is what we are aiming at in this section, foam evolution is essential to comprehend the basic mechanisms such as nucleation, growth, coalescence and – last but not least – foam stabilization.[9]

3.2.1 Cellular Structure

3.2.1.1 Cell Structure Evolution

Different foam evolution states are passed through when starting at the compact skin near the mold wall and moving in direction of the center of the casting.[10] Figure 3.12 shows density profiles and cell structures at different relative densities for two aluminum castings produced by HP-IFM.

The foam morphology reflects the competition between decomposition of the blowing agent and solidification. The structures appear strongly jagged at the steep transition between the compact skin and the cellular core. These kinds of structures are characteristic for the presence of a very high viscosity during the development of the porosity. Comparing the experimental finding with numerical simulation, see Fig. 5.3, the viscosity is estimated to be by a factor of 10^5 larger than that of the liquid metal. Combining this result with the viscosity of semisolid metal slurries,

[8] We have already first experimental evidence that blowing agent decomposition is partially suppressed if the the dwell pressure is too high.

[9] See Sect. 4.3 for foam stabilization mechanisms.

[10] The foam evolution time increases with increasing distance from the mold wall. Thus, foam evolution may be investigated by means of a single sample.

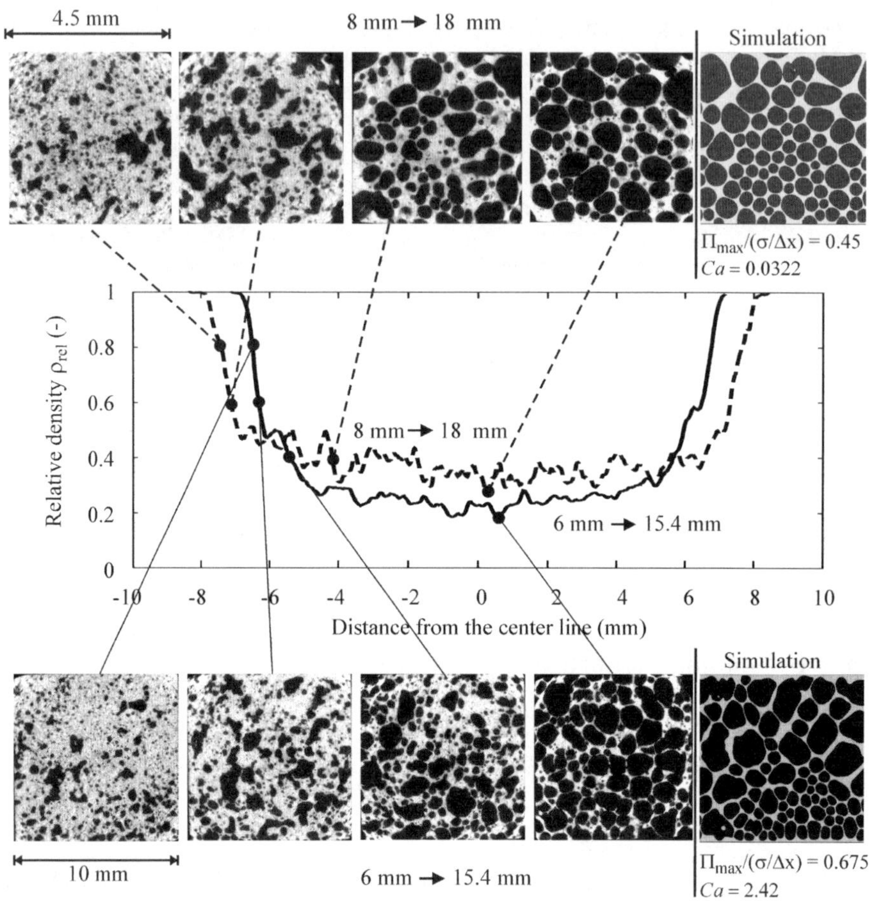

Figure 3.12: Cell structure evolution for different relative densities for two AlSi9Cu3 HP-IFM castings. *Upper row*: 8 mm → 18 mm. *Lower row*: 6 mm → 15.4 mm. (μCT, simulation: see Chap. 5, Fig. 5.9.)

see Fig. 4.11, it becomes evident that the porosity develops just before complete solidification of the metal takes place.

The two examples depicted in Fig. 3.12 show different total heat content. As a result, the cells of the example of the upper row (8 mm → 18 mm) appear somewhat rounder than that of the lower row (6 mm → 15.4 mm). Along the more or less pronounced density plateau of the core, the cell structure does not change very much.

Trying to reproduce the cell structures of the density plateaus by means of numerical simulation is becomes evident that a high viscosity is necessary but not sufficient to obtain the observed structures. Coherence between experiment and simulation is achieved not before an additional stabilizing force, a disjoining pressure Π, is postulated.[11] The disjoining pressure allows the development of stable, straight cell walls. If the viscosity is not too high (upper row), the disjoining pressure prevents strong cell coarsening while cell rearrangement and spheroidization are still possible. As a result, the structure shows quite round cells and pronounced straight cell walls and Plateau borders. If the viscosity is by a factor of 100 larger (lower row), the disjoining pressure still stabilizes cell walls but the viscous forces partly prevent spheroidization of the bubbles. As a result, the cells appear more angular.

Any kind of foam stabilization in addition to the dynamic stabilization due to a high viscosity is kind of unexpected since it is well known that pure liquids do not foam.[12] In the following, more experimental evidence for the presence of stabilization by a static force is collected.

3.2.1.2 Cell Wall Stabilization

Figure 3.13 shows a section and local magnification of the cell structure of a HP-IFM part. All-over the section, straight and very thin cell walls are present. The magnification reveals a typical polygonal foam structure. Although numerical simulation also predicts the formation of some straight cell walls for high viscosities (see Fig. 5.3), the development of polygonal cell structures has never been observed without any additional stabilizing force.

3.2.1.3 PM foam vs IFM Foam Structure

In order to assess the degree of foam stabilization it is instructive to compare IFM cell structures with that of PM foam structures of similar relative density, see Fig. 3.14.

The resulting cell structures of the PM foam and the HP-IFM foam are amazingly similar. In both cases, an ideal cell structure does not develop. The cells are deformed and there are odd cell configurations present.

PM foams are generally regarded as to be stabilized by oxides provided by the underlying metal powder. These oxides are not present in form of solid particles but form network-like, loose structures which act as mechanical barriers in cell walls, see Fig. 4.14. This barrier effect is the origin of the observed stabilization [4,45,146]. We

[11] For details, see Sect. 4.3.2.

[12] For a detailed discussion see Chap. 6.

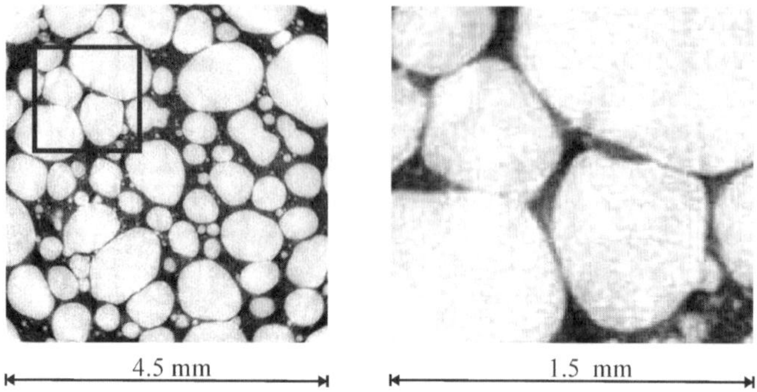

Figure 3.13: Evidence for cell wall stabilization. The magnification reveals the presence of polygonal cells with straight, very thin cell walls. (μCT, HP-IFM, 8 mm → 18 mm, dark: AlSi9Cu3)

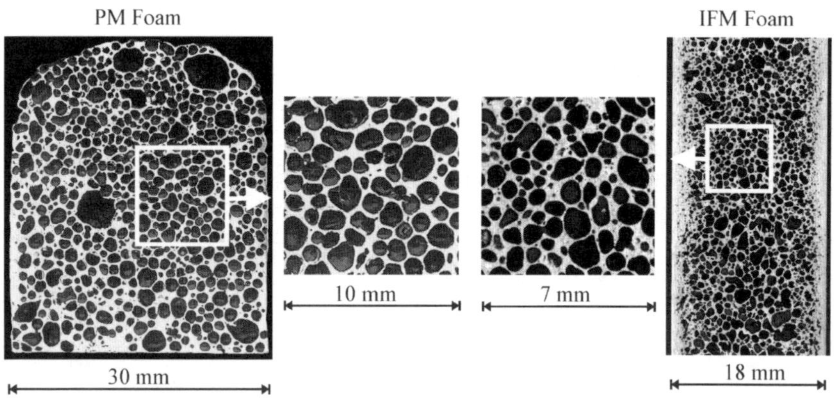

Figure 3.14: PM foam structure vs IFM foam structure. *Left*: PM foam (Al powder + TiH_2, hot pressed, furnace heating). *Right*: HP-IFM (8 mm → 18 mm, AlSi9Cu3).

suppose that integral foams are stabilized in a very similar way [149].[13] Solidification and foam formation take place simultaneously. As a result, the solid phase particles of the semi-liquid metal slurry act as mechanical barriers analogous to the oxide structures in PM foams. Consequently, the cell structures of PM and IFM foams appear very similar.

[13]For theoretical details, see Chap. 6.

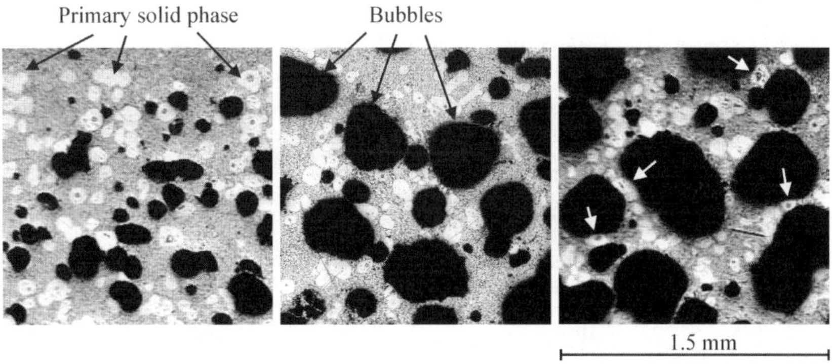

Figure 3.15: Stabilizing effect of the primary solid phase (bright) during thixomolding. From *left* to *right*: Phase ratio $\phi = \frac{V_G}{V_F}$ increases (V_G and V_F denote the total gas and fluid volume). *Left* picture: Bubble coalescence is not prevented by the particles. *Right* picture: Primary solid phase particles get captured between the bubbles and prevent further cell wall thinning (*white arrows*). (light microscopy, LP-IFM, thixomolding, AZ91, relative volume fraction of primary solid phase $V_{rel} \approx 0.3$)

3.2.1.4 Mechanical Barrier Effect

How can we make the barrier effect of solid phase particles visible? The observation of the barrier effect of solid phase particles is possible by exploring thixomolding when working in the semi-liquid state. In this case, there are two classes of solid particles present during IFM, large and small ones. The solid phase particles already present in the semi-liquid melt before injection have a mean diameter of about 100 µm (= primary solid particles) whereas the particles which develop during solidification are by a factor of ten smaller.

Figure 3.15 shows the stabilizing effect of the primary solid particles. For low phase ratios ϕ, the solid phase particles with a quite high relative volume fraction of about $V_{rel} \approx 0.3$ do not hamper coalescence very much. They are expelled from cell walls and have only a minor effect on cell wall stabilization. Things are different for higher phase ratios. In this case, the particles get captured between two bubbles and represent mechanical barriers against coalescence. The resulting cell wall thickness is about the particle diameter.[14] The solid phase particles which develop during solidification of the casting are assumed to show the same stabilizing effect as the primary solid phase but on a much smaller length scale.[15]

[14]The analogous numerical result is depicted in Fig. 6.9.

[15]This is the reason why it is not reasonable to use semi-liquid melts for IFM. In addition, the lower heat content of semi-liquid melts is detrimental for mold filling and foam formation. Thus, the specific property of thixomolding to process semi-liquid melts is unfavorable for IFM.

3.2.2 Evolution Laws

In the last section, the appearance of the cell structure was employed to find hints for foam stabilization during IFM. Now, we use the evolution of the cell structure, especially the relation between the mean cell diameter D and the phase ratio ϕ, to deduce more information about the underlying foam formation and stabilization mechanisms. In Part II, Chap. 5 we show that $D(\phi)$ reveals much about the basic phenomena which are important during foam formation. Here, we use these findings to interpret the experimental results.

Figure 3.16 shows that there is a strong correlation between the mean cell diameter D and the relative density ρ_{rel}.[16]

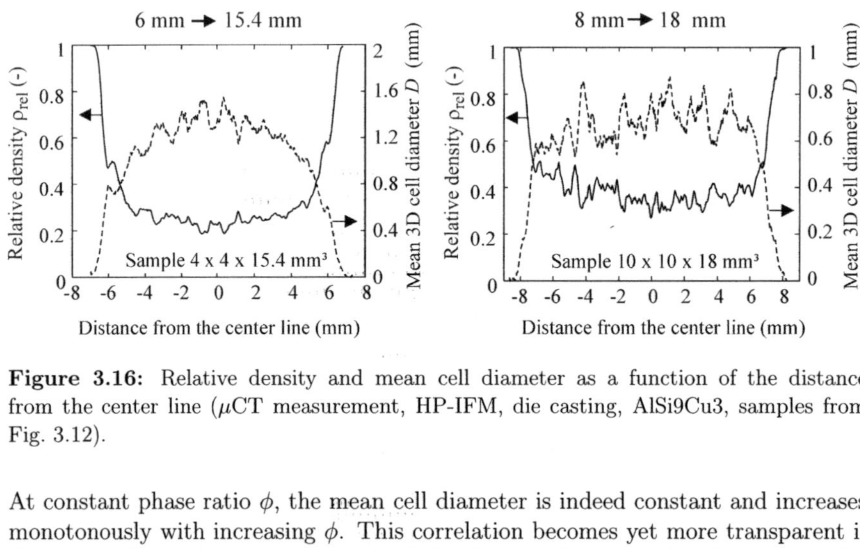

Figure 3.16: Relative density and mean cell diameter as a function of the distance from the center line (μCT measurement, HP-IFM, die casting, AlSi9Cu3, samples from Fig. 3.12).

At constant phase ratio ϕ, the mean cell diameter is indeed constant and increases monotonously with increasing ϕ. This correlation becomes yet more transparent if the mean cell diameter as a function of the phase ratio is considered, see Fig. 3.17. The logarithmic plot reveals different regions characterized by different exponents n which we refer to as evolution exponents.[17] In the following, we apply the realizations derived from numerical experiments, Chaps. 5 and 6, to identify the basic mechanisms which govern integral foam structure evolution.

At low phase ratios, foam evolution is experimentally determined to follow

$$D \propto \phi^{1/3}. \tag{3.6}$$

[16] The determination of the mean cell diameter is described in Appendix B.2.

[17] The evolution exponent is defined in Eq. (5.1).

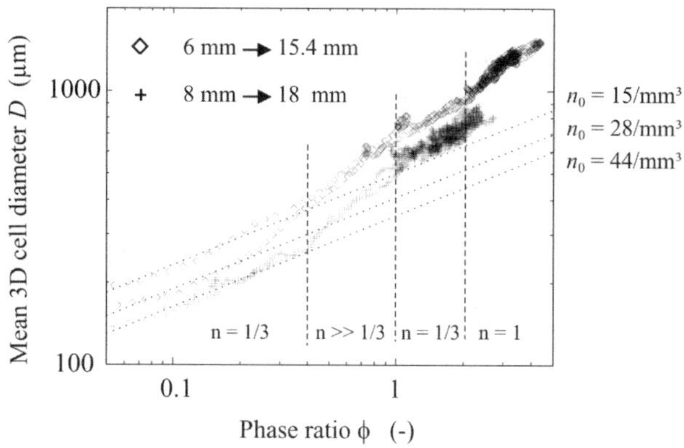

Figure 3.17: Mean 3D cell diameter D as a function of the phase ratio ϕ. Dotted lines: Coalescence-free evolution with different nuclei densities n_0, Eq. (3.7). (μCT measurement, HP-IFM, die casting, AlSi9Cu3, samples from Fig. 3.12)

The exponent 1/3 is characteristic for coalescence-free expansion, compare Eq. (5.22):

$$D = \sqrt[3]{\frac{6}{n_0\,\pi}\,\phi}, \tag{3.7}$$

where n_0 denotes the nuclei density. Thus, the initial nuclei density can be deduced from the curve progression in Fig. 3.17. We find nuclei densities in the range between $10/\text{mm}^3$ and $100/\text{mm}^3$.

The experiments show that deviations from coalescence-free behavior, which indicate the onset of bubble coalescence events, start between $\phi = 0.3$ and $\phi = 0.4$. The onset of coalescences reveals that foam evolution is – as anticipated – non-stabilized in this regime, compare Sect. 5.2.2. From theory, the mean cell diameter is expected to follow non-stabilized evolution with a divergence at about $\phi \approx 4$ with increasing ϕ, see Eq. (5.23). In fact, the mean cell diameter starts to increase strongly for $\phi > 0.4$.

Surprisingly, non-stabilized evolution stops at about $\phi = 1$ and there is no indication of any kind of divergence. On the contrary, the curve progression for higher phase ratios shows again an exponent of $n = 1/3$, i. e. coalescence-free evolution, which eventually changes to $n = 1$. Thus, instead of the expected non-stabilized cell growth we find coalescence-free growth which eventually changes to an evolution behavior characterized by an evolution exponent $n = 1$ (sample 6 mm $\rightarrow$ 15.4 mm). The evolution exponent $n = 1$ describes foam evolution in the presence of a bounded above stabilizing disjoining pressure, see Eq. (5.24). Summarizing, there is strong evidence for the presence of some kind of foam stabilization mechanism during IFM.

From theory, the mean cell diameter during coalescence-free evolution as well as non-stabilized evolution are predicted to be governed by $1/\sqrt[3]{n_0}$. It is thus instructive to plot the mean cell diameter normalized by $1/\sqrt[3]{n_0}$, see Fig. 3.18.[18] After normalization, the data of the two experiments lie – as predicted by theory (Chap. 5) – on the same curve!

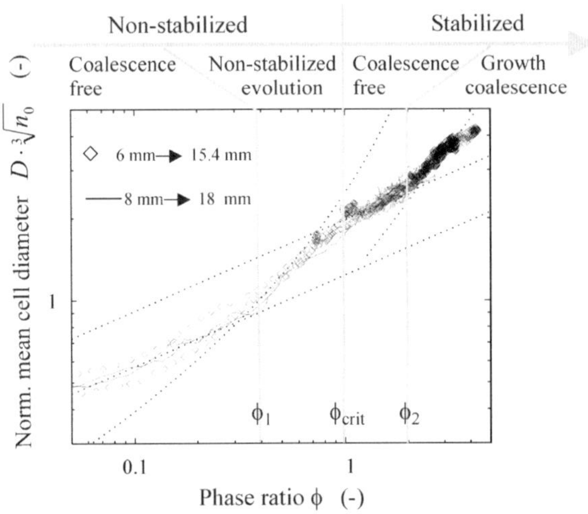

Figure 3.18: Evolution regimes for integral foam. Normalized mean 3D cell diameter D as a function of the phase ratio ϕ. Foam evolution appears to be non-stabilized for $\phi < \phi_{\mathrm{crit}}$ and stabilized for $\phi > \phi_{\mathrm{crit}}$. ($\mu$CT measurement, HP-IFM, die casting, AlSi9Cu3, samples from Fig. 3.12, *dotted lines*: analytical solutions)

At first glance, the experimental finding appears rather strange since non-stabilized evolution changes to stabilized evolution during expansion. What kind of effect is able to explain that stabilization is not present at low phase ratios but emerges for higher ones? The mechanical barrier effect of solid phase particles shown in Fig. 3.15 already points in the right direction. Numerical simulation is crucial to fully understand the underlying mechanism. We suppose that the stabilization mechanism is based on the semi-liquid melt present during foaming. The effect of the solid phase particles is investigated in Chap. 6 where the expansion behavior of a fluid containing particles which are wet by the fluid is considered. During foaming, the particles get confined in cell walls thereby stabilizing them by inducing

[18] This normalization compares the mean cell diameter with the mean bubble nucleus distance $1/\sqrt[3]{n_0}$.

a stabilizing force. A very surprising numerical result is, that the stabilizing effect of the particles depends on the phase ratio. For low phase ratios, the particles are still able to escape from evolving cell walls due to the yet high mobility and their stabilizing effect is rather weak. The situation changes for a higher phase ratio. In this case, the particles get trapped in cell walls and start to act as barriers thereby inducing a stabilizing counter pressure against cell wall drainage and rupture, see also Fig. 3.15.

Summarizing, the following expansion regimes are identified by the experiment (Fig. 3.18) and confirmed by numerical simulation (Chap. 6):

1. Non-stabilized evolution
   - $\phi < \phi_1$: Coalescence-free expansion with exponent $n = 1/3$. Coalescence does not take place since the distance between the bubble nuclei is still high.
   - $\phi_1 < \phi < \phi_{\mathrm{crit}}$: Non-stabilized evolution with exponent $n \gg 1/3$.
2. Stabilized evolution
   - $\phi_{\mathrm{crit}} < \phi < \phi_2$: Stabilized evolution without coalescence and exponent $n = 1/3$. Stabilization is sufficient to suppress bubble coalescence.
   - $\phi > \phi_2$: Stabilized evolution with $n = 1$. Evolution is accompanied by coalescence events (growth coalescence).

The four regimes are evident in Fig. 3.18. Especially the transition between non-stabilized and stabilized expansion at $\phi_{\mathrm{crit}} \approx 1$ is a clear proof for the presence of stabilization during IFM.

All evolution regimes except stabilized evolution with exponent $n = 1$ are governed by $1/\sqrt[3]{n_0}$. Finer cell structures can only be achieved by increasing n_0 as long as $\phi < \phi_2$. Generally, the nuclei density is very important in view of a fine and homogeneous cell structure. What determines n_0 and how can n_0 be influenced?

3.2.2.1 Nucleation

Nucleation is a key issue during IFM. It seems likely that the blowing agent particles not only act as gas supply but also have the function of nucleation sites [41]. On the one hand, nucleation is facilitated at particles. On the other hand, the hydrogen concentration shows pronounced peaks at blowing agent particles on the relevant time scale, see Sect. 4.1.2.2. In fact, blowing agent particles are very often observed to be located at bubble nuclei, see Fig. 3.19.

This observation is further supported by a simple estimate of the number of blowing agent particles contained within one mm^3 of metal. We assume that the

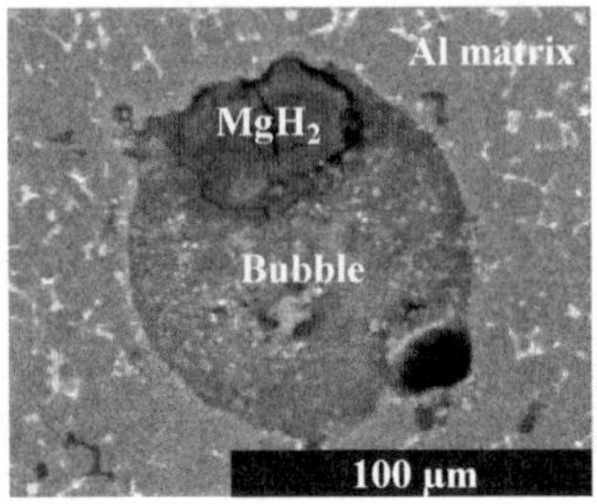

Figure 3.19: Nucleation at MgH_2 particle. (LP-IFM, die casting, AlSi9Cu3)

particles are disc-like with a mean radius of 50 µm and a mean thickness of 20 µm.[19] A typical particle volume fraction of 0.01–0.02 is thus equivalent to a nuclei density of about $50/\text{mm}^3$–$100/\text{mm}^3$. These nuclei densities coincide with the experimental findings. The blowing agent particles may thus be assumed to act as nucleation sites. Consequently, the nuclei density can be increased by decreasing the blowing agent particle size while maintaining the particle volume fraction.

3.3 Mechanical Properties

Integral foams are structures not materials! The characterization of integral foam is always based on structures, generally on sandwich structures. The experimentally determined properties are structure properties and not material constants.

Integral foams show outstanding properties with respect to bending stiffness, energy absorption and damping capability. These features are a direct consequence of the foamed core which either may simply act as space holder or may be successively compressed in case of energy absorption. The basic mechanisms which determine the bending stiffness and the energy absorption capability are more or less well understood. However, damping of metal foam is still obscure and still a matter of discussion.

3.3.1 Bending Stiffness of Sandwich Beams

The weight specific elastic modulus of metal foam is always inferior to that of compact material, see Eq. (1.5). Nevertheless, in bending, foam is superior to compact material due to the principle of mass separation. This effect is even further pro-

[19]See Fig. 3.6.

nounced if sandwich beams are considered where the primary role of the foamed core is to maintain the distance between the solid face sheets.

3.3.1.1 Effective Elastic Modulus

The bending stiffness of integral foam is described with the help of sandwich beams characterized by a continuous transition between the compact skins and the cellular core, see Fig. 3.20, left.

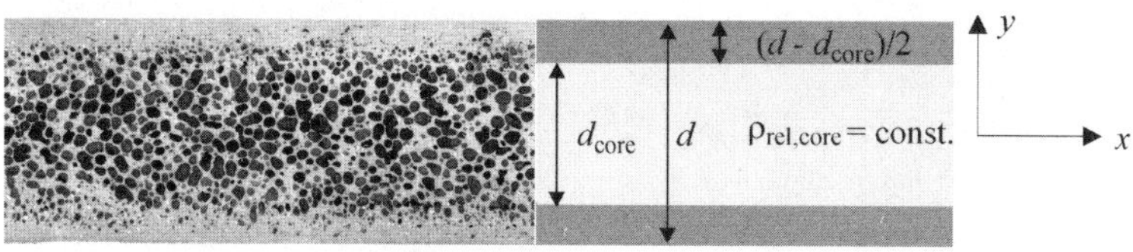

Figure 3.20: Integral foam structure (*left*) and sandwich model (*right*) with uniform core and dense skin (d: sandwich thickness, d_{core}: core thickness, $(d - d_{\mathrm{core}})/2$: skin thickness).

The maximum deflection for 3-point-bending, $\delta_{\max}$, is given by

$$\delta_{\max} = \frac{F\, l^3}{48\, B_{\mathrm{F}}}, \tag{3.8}$$

with the applied load F and the span l. The **beam stiffness**, B_{F}, is defined as

$$B_{\mathrm{F}} = 2\, b \int_0^{d/2} E_{\mathrm{F}}(y)\; y^2 dy, \tag{3.9}$$

where b designates the width of the beam and $E_{\mathrm{F}}(y)$ is the local elastic modulus of the beam. From Eq. (3.9) it is evident that the resulting beam stiffness depends on the local elastic properties which are determined by the material distribution. The beam stiffness of a compact beam, B, is given by

$$B = \frac{1}{12}\, b\, d^3\, E = I\; E, \tag{3.10}$$

where I denotes the moment of inertia and E the elastic modulus of the compact material.

The determination of the elastic modulus of an integral foam beam, e. g. by impulse excitation technique, is always based on the assumption that the structure is homogeneous. Thus, the measured elastic modulus is an **effective modulus** E_{eff} of the foam structure given by

$$E_{\mathrm{eff}} = \frac{B_{\mathrm{F}}}{I} = \frac{24}{d^3} \int_0^{d/2} E_{\mathrm{F}}(y)\, y^2\, dy. \tag{3.11}$$

Equation (3.11) clearly demonstrates that the experimentally determined elastic modulus is a complex mean value of the local elastic properties of the beam. The relative effective modulus, E_{eff}/E, is identified with the relative beam stiffness, B_{F}/B:

$$\frac{E_{\mathrm{eff}}}{E} = \frac{B_{\mathrm{F}}}{B}. \tag{3.12}$$

Figure 3.21 shows the relative effective modulus of magnesium sandwich beams as a function of the mean relative density.[20] The measured relative effective elastic

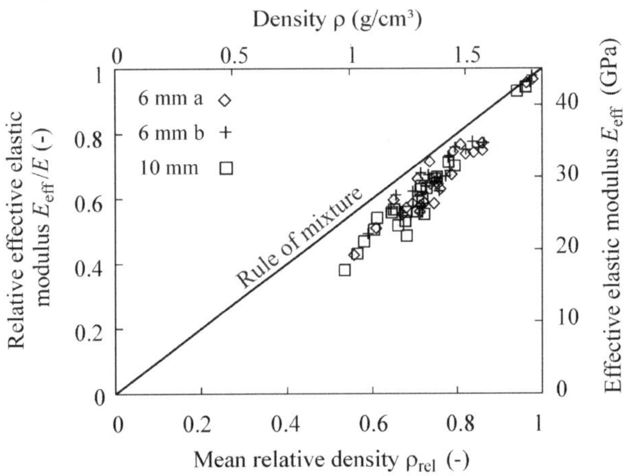

Figure 3.21: Relative effective elastic modulus as a function of the mean relative density for d = 6 mm and d = 10 mm sandwich beams [113, 150]. *Full line*: Rule of mixture. (LP-IFM, thixomolding, magnesium alloy AZ91, a and b denote different positions of the cast part. Sample dimensions: $6 \times 6 \times 70$ mm^3 and $9.5 \times 10 \times 98$ mm^3)

modulus lies below the rule of mixture. This result was not expected since the density of the beams is reduced by removing material from the core which makes a minor contribution to the beam stiffness compared to the skin regions, see Eq. (3.11). Interestingly, polymer integral foams show an analogous behavior [63, 111, 210].

3.3.1.2 Mechanical Model

In order to achieve a better understanding of the stiffness of integral foam structures a simple sandwich model is applied.[21] The real structure is replaced by an idealized sandwich with compact skins and uniform core density ρ_{F}, see Fig. 3.20, right.

[20] The elastic modulus of integral foam beams is determined by impulse excitation technique according to ASTM E 1876-01: Standard test method for dynamic Young's Modulus, Shear Modulus, and Poisson's Ratio by impulse excitation of vibration.

[21] For details, see N.C. Hilyard: *Mechanics of cellular plastics* [111]. See also [185].

To calculate B_F, the elastic modulus of the foamed core $E_F = E_F(\rho_{rel,core}(y))$ has to be known. There are plenty of different models which try to describe the dependence of the elastic modulus of foam on its relative density [85, 111], see also Eq. (1.5). These sometimes complex models show no advantage compared to the simplest relationship available, the **power-law-model**,

$$E_F(\rho_{rel,core}(y)) = E \cdot \rho^n_{rel,core}(y), \tag{3.13}$$

where E denotes the elastic modulus of the compact material and n the power law exponent with typically $1 \leq n \leq 2$.
The elastic modulus is thus modeled as:

$$E(y) = \begin{cases} E\,\rho^n_{rel,core} & \text{for} \quad |y| \leq d_{core}/2 \\ E & \quad |y| > d_{core}/2. \end{cases} \tag{3.14}$$

The relative core density, $\rho_{rel,core}$, can be expressed as a function of the relative beam density

$$\rho_{rel,core} = \frac{\rho_{rel} - \epsilon}{1 - \epsilon}, \tag{3.15}$$

where the skin fraction ϵ is given by $\epsilon = (d - d_{core})/d$.
Combining Eqs. (3.9), (3.10), (3.14) and (3.15) the **relative bending beam stiffness** may be expressed as [111]:

$$\frac{B_F}{B} = \left(\frac{\rho_{rel} - \epsilon}{1 - \epsilon}\right)^n (1 - \epsilon)^3 + \left[1 - (1 - \epsilon)^3\right] = \frac{E_{eff}}{E}. \tag{3.16}$$

Figure 3.22 shows the curve progressions predicted by Eq. (3.16) for different skin fractions and $n = 2$. If the skin thickness is not too high, the effective elastic modulus lies below the rule of mixture for high densities. This is a consequence of the quadratic dependence of the elastic modulus on the relative density. With decreasing relative density the stiffness eventually crosses the rule of mixture only for thick skins (large ϵ). Sandwiches with thin skins always remain below the rule of mixture.

Experimental data and the theoretical model are compared in Fig. 3.23. The theoretical model predicts the effective elastic modulus to decrease with increasing sandwich thickness if the skin thickness is constant. In fact, the data for the 10 mm beams is somewhat below that for the 6 mm beams. A rather thin skin of 0.5 mm has to be assumed in order to match the theoretical prediction with the experiment. Whether this result is an indication that the model is not suitable is not clear at the moment. A good reason for the low skin thickness could be that not the mean ($\approx$ 1 mm) but the minimal skin thickness is important.

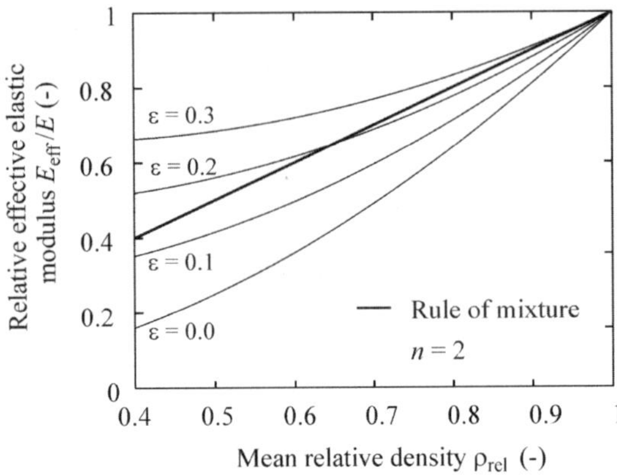

Figure 3.22: Relative effective elastic modulus of sandwich beams for different skin fractions ϵ (Eq. (3.16) for $n = 2$). *Full bold line*: Rule of mixture.

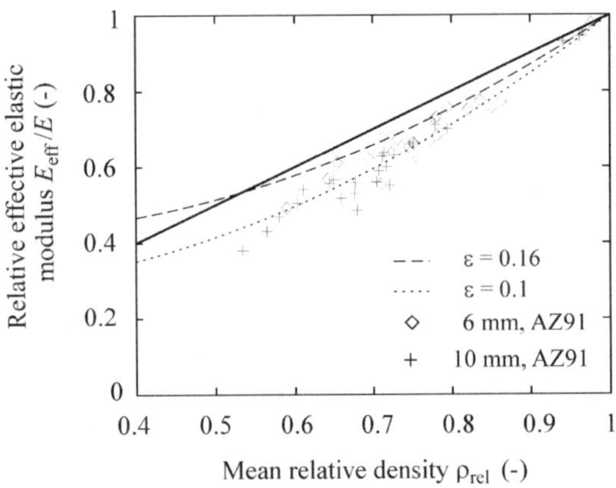

Figure 3.23: Relative effective elastic modulus of sandwich beams: Experiment (Fig. 3.21) vs theory (Eq. (3.16) for $n = 2$). The skin thickness is assumed to be 0.5 mm.

3.3.2 Compression Behavior

As already mentioned, integral foam parts are not homogeneous materials but structures. The outcome of the measurement of stress–strain curves are thus structure properties and not material properties.

In the following, the compression behavior of wall material from integral foam castings is considered. Strain–stress curves of magnesium LP-IFM sandwich structures with different mean densities are depicted in Fig. 3.24. The tested samples

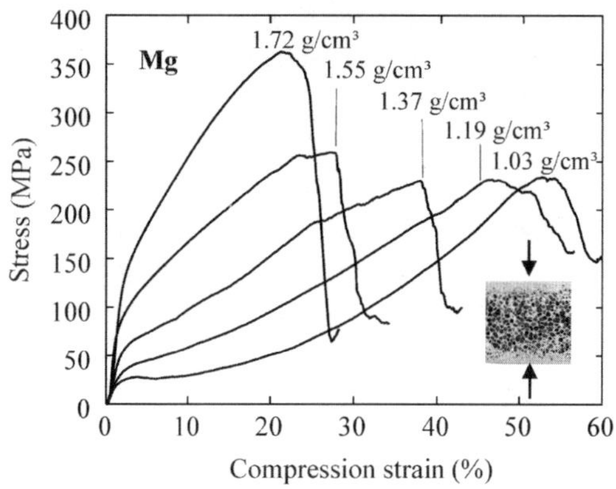

Figure 3.24: Stress–strain curves for magnesium LP-IFM sandwich structures with different mean density [113, 150]. With decreasing density a stress plateau appears which is typical for foam structures. (force perpendicular to the dense skin, test velocity: 0.4 mm/min, sample dimension: $6 \times 6 \times 5.8$ mm^3)

show a sandwich structure with load direction perpendicular to the solid skin. Thus, the foamed core of the casting is characterized by this test. The stress plateau, which is typical for foamed material (see Fig. 1.7) [13], emerges with decreasing density.

The compression strength is determined by the minimum density of the core. Generally, the foam strength follows a power law relation (Eq. (1.6)) and decreases strongly with decreasing core density. Hence, not the mean density but the minimum density of the core determines the yield strength. The further incline of the stress–strain curve is governed by the density distribution of the core and the onset of compaction. A plateau of the density distribution leads to a similar plateau of the stress–strain curve.

Integral foam castings are completely surrounded by a solid skin. The mechanics of these structures is determined by the geometry of the whole casting and can not be predicted easily.

3.3.3 Damping Behavior

Generally, damping of foam is always a structure property and never, even if homogeneous foam is considered, a material property.[22] In any case, the damping behavior of a foam part or an integral foam casting is a structure property. This is also true if damping is determined with the help of standardized samples.

In the following, damping of integral sandwich beams at low amplitudes is investigated. Figure 3.25 shows the internal friction[23] Q^{-1} of magnesium and aluminum integral sandwich beams as a function of the porosity [152, 153].

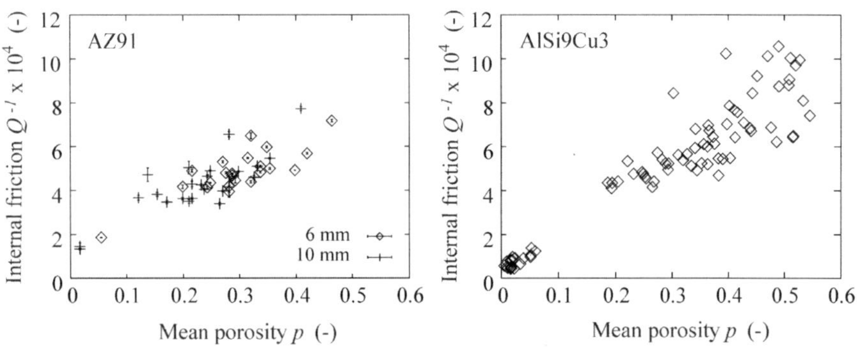

Figure 3.25: Internal friction of integral foam sandwich beams as a function of the mean porosity. *Left*: Magnesium LP-IFM sandwich beams (sample dimension: $6 \times 6 \times 70$ mm^3, $f \approx 4500$ Hz and $10 \times 10 \times 98$ mm^3, $f \approx 4500$ Hz). *Right*: Aluminum HP-IFM sandwich beams (sample dimension: thickness from 8 to 18 mm, $f = 3000 - 8000$ Hz).

The resonance frequency depends on the sample geometry and the material properties. It is about 4500 Hz for the magnesium samples and between 3000 and 8000 Hz for the aluminum samples in Fig. 3.25. For aluminum as well as for magnesium, the internal friction increases linearly with the porosity p,

$$Q^{-1}(p) = Q^{-1}(0) + a \cdot p, \tag{3.17}$$

where a is constant.

[22] Damping is strongly dependent on the resonance frequencies which are determined by the geometry. In addition, the geometry has strong influence on thermoelastic damping or the stress–strain distribution in the foam part.

[23] The internal friction, see Eq. (1.7), is determined from damped oscillations measured by impulse excitation technique according to ASTM E 1876-01: Standard test method for dynamic Young's Modulus, Shear Modulus, and Poisson's Ratio by impulse excitation of vibration.

Although there is a clear correlation between internal friction and mean porosity, which is also reported in [97, 127, 171, 172], Q^{-1} still shows strong variations at constant porosity. A possible explanation for this behavior might be that damping is dependent on the mean cell diameter as proposed in [171]. In addition, the resonance frequency and the beam thickness may have influence on the damping behavior. Also inhomogeneous material distributions might explain these variations.

Altogether, damping of metal foam is still obscure due to the variety of external and internal parameters which have influence on the working mechanisms. Generally, the internal friction is dominated by the porosity and is by a factor 10–20 larger compared to compact material for a relative density of about 0.5, see Fig. 3.25. At the moment, a direct transfer of results obtained with the help of test geometries to real castings is not possible. However, first investigations of the damping behavior of complex integral foam parts show a similar increase of the internal friction as that of test samples.

3.4 Synopsis

The decomposition dynamics of the blowing agent and the foaming pressure are essential to understand IFM:

- The decomposition kinetics of the blowing agent determines the density profile.
- The dense skin is the result of the chilling effect of the mold walls and the delayed decomposition of the blowing agent.
- The shape of the density profile reveals the degree of decomposition of the blowing agent. The appearance of a plateau in the casting is an indication for complete decomposition.
- During LP-IFM, the relative density may be controlled by the dwell pressure which balances the foaming pressure.
- The foaming pressures are quite high, about 10^2 bar.

The investigation of the cell morphology reveals much about the foam formation mechanisms which are not directly accessible by experimental observation:

- Cell nucleation takes place at blowing agent particles. The nuclei density n_0 is thus determined by the blowing agent particle density. n_0 can be deduced from the foam evolution behavior at low phase ratios.
- Cell diameter and relative density are intimately correlated.

- Integral foam is stabilized by the solid phase of the semi-liquid metal. The evolution behavior of integral foam is analogous to that of liquids containing particles. This observation is a fundamental result and has very important implications:
  - Very low relative densities may be realized without severe coarsening.
  - Melt conditioning is not necessary.
  - Recycling is analogous to standard die castings.
  - The solidification behavior of the casting alloy is expected to be essential. The minimal cell wall thickness is determined by the grain diameter of the solidification structure, see also Chap. 6.
- At the beginning of expansion, the mean cell diameter is proportional to $1/\sqrt[3]{n_0}$. At large phase ratios, the mean cell diameter becomes independent from the initial nuclei density.

The properties of integral foam are always structure properties:

- The properties are dominated by the relative density.
- As a rule of thumb, the bending stiffness of sandwich beams is just below the rule of mixture.
- The internal friction increases linearly with decreasing relative density and is for $\rho_{\mathrm{rel}} = 0.5$ by a factor of 10–20 higher than that of dense material.
- Aluminum and magnesium integral foams show analogous behavior.

Part II
PHYSICS

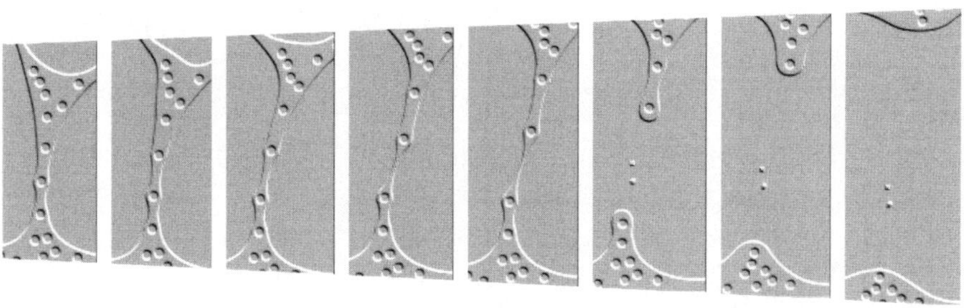

Die Physik ist für die Physiker eigentlich viel zu schwer.
David Hilbert

Integral foam molding as proposed in Chap. 2 raises fundamental questions:

- Is foam evolution possible within the available foaming time of a fraction of a second or hampered by inertia, viscous or capillary forces?
- What is the role of gas diffusion? Is there a possibility to compensate inhomogeneities of the distribution of the blowing agent particles by diffusion?
- Might foam formation be hampered by the gas solubility of the metal?
- How is – if at all – foam stabilization realized? For all known metal foaming processes conditioned melts have to be used in order to stabilize the foam structure.
- What are the main structure characteristics of the resultant cell structures?
- Is there – analogous to well-known foaming processes – an intimate relation between mean cell size and relative density?
- What is the influence of the initial bubble nuclei density?

In order to find answers to these questions we have to have a closer look on the fundamental physical foaming mechanisms with special emphasis on the actual situation during IFM.

The first chapter of this part is a short introduction into the basic governing equations, bubble dynamics and foam stabilization mechanisms, especially the stabilization of metal foam.

The second chapter is devoted to generally valid foam evolution laws. It is shown that the evolution of the cellular structure is determined by very simple rules. These rules allow to extract important information from experimental results about the presence or absence of foam stabilization. In addition, the role of the initial bubble nuclei density on the foam structure is revealed.

The last chapter is concerned with foam stabilization during IFM. The grain nuclei present in the melt during endogenous solidification of the casting act as obstacles in cell walls. This obstacle effect generates a stabilizing force, the disjoining pressure, which is responsible for foam stabilization and the characteristic foam structure.

The derivation of the evolution laws and the discussion of foam stabilization by particle confinement is based on analytical and numerical approaches. The details of the numerical approach, which would be hindering here, are presented in Part III.

4 Physics of Foaming

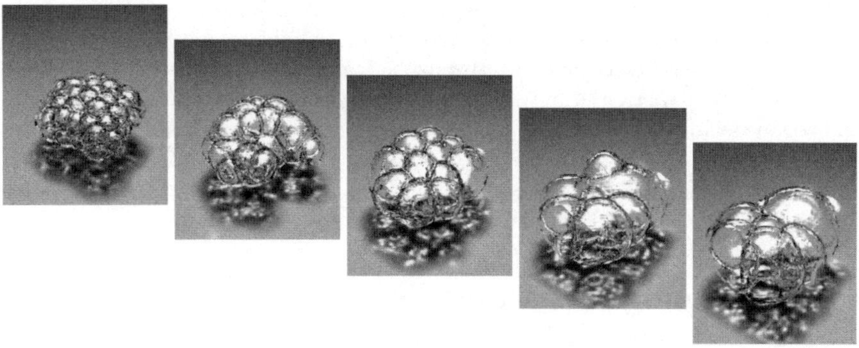

As far as the laws of mathematics refer to reality, they are not certain; and as far they are certain, they do not refer to reality.

Albert Einstein

Foaming by in situ gas generation is a growth process where different evolution stages are passed through, see Fig. 4.1. At the very beginning, expansion starts

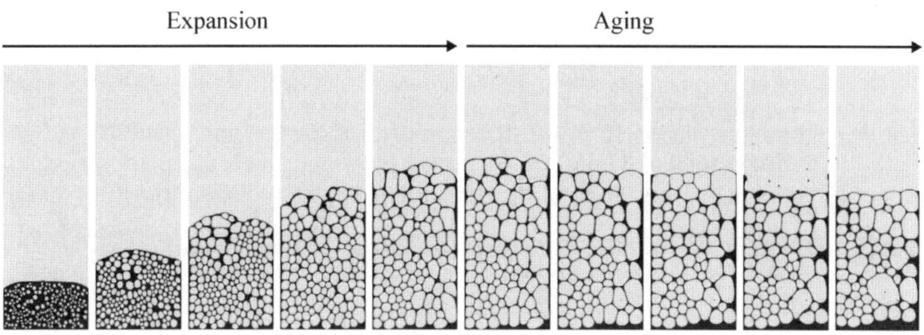

Figure 4.1: Evolution stages of a foam: Nucleation, growth, coarsening and decay. (2D-LBM Simulation)

with bubble nucleation induced by excessive supply of a blowing gas. This gas is released and dissolved in the liquid phase due to a temperature increase or an external pressure drop, which both cause the blowing agent to decompose thereby increasing the concentration of dissolved gas in the fluid. As a result, gas diffuses to the bubbles, the amount of gas in the bubbles increases and makes them grow. During growth, the bubbles start to interact with each other. At the beginning, the forces which the bubbles mutually exert on each other result from hydrodynamics and are purely viscous and hence dependent on the expansion velocity and the viscosity of the fluid. These forces try to deform the bubbles and are counteracted by the surface tension seeking to restore spherical bubbles. If the relative density reaches a certain value, the bubbles transform to components of a foam structure and assume a more and more polygonal shape. This process is intimately linked with cell coalescence which leads to foam coarsening and eventually to foam collapse.

The aim of this chapter is to discuss the governing physical laws of foam evolution with special emphasis on the inherent time scale of IFM.

4.1 Governing Equations

The development of foam underlies basic physical laws which are briefly described. Foam is a two phase system consisting of a gas and a fluid.[1] Due to the large density difference between gas and fluid, the dynamics of the gas is insignificant and can be neglected for the most practical applications.

4.1.1 Hydrodynamics

The dynamics of an incompressible fluid is given by the famous incompressible **Navier-Stokes equations (NSE)**:[2]

$$\partial_\alpha v_\alpha = 0 \tag{4.1}$$

$$\partial_t v_\alpha + (v_\beta \, \partial_\beta) \, v_\alpha = -\frac{1}{\rho}\partial_\alpha p + \nu \, \partial_\beta^2 v_\alpha + g_\alpha \tag{4.2}$$

with density ρ, velocity $\mathbf{v}$, pressure p, kinematic viscosity ν and gravitational acceleration $\mathbf{g}$. Equations (4.1) and (4.2) express mass and momentum conservation, respectively. Solving the NSE gives the pressure and velocity field of the fluid phase.

The gas pressure p_i in bubble i is given by the **ideal gas equation**:

$$p_i = \frac{n_i \cdot R \cdot T}{V_i} \tag{4.3}$$

[1] IFM is actually involved with three phases, gas, fluid and solid.

[2] Einstein's summation rule is employed: $x_\alpha x_\alpha \equiv \sum_\alpha x_\alpha x_\alpha$.

where R denotes the ideal gas constant and n_i the amount of gas, T the temperature and V_i the volume of bubble i.

The coupling between gas and liquid via momentum transfer takes place at the fluid–gas interface denoted as Γ. At the interface, the velocity of gas and fluid have to be equal:

$$\mathbf{v}_{\mathrm{G}}(\mathbf{x}) = \mathbf{v}_{\mathrm{F}}(\mathbf{x}) \qquad \forall \mathbf{x} \in \Gamma. \tag{4.4}$$

In addition, force equilibrium has to be maintained, i. e. the force exerted from the gas on the liquid has to be the same as that exerted from the liquid on the gas, see Fig. 4.2. Since a gas is only able to exert normal forces but can not transfer tangential stresses, the **boundary conditions** split up into a normal and a tangential component,

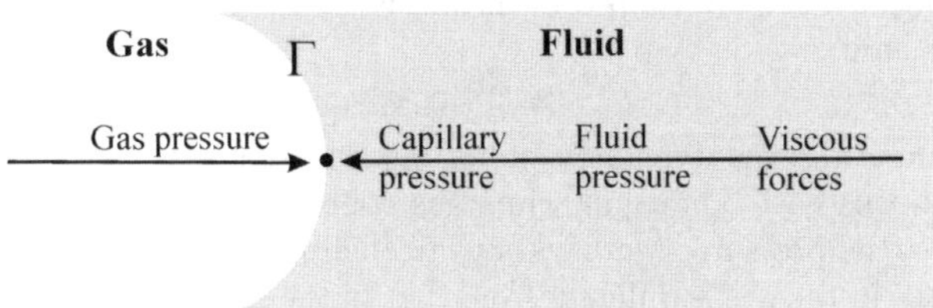

Figure 4.2: Force balance at the fluid–gas interface. The gas pressure is locally balanced by the capillary pressure, the fluid pressure and the viscous forces.

$$p - 2\,\rho\,\nu\,\partial_n v_n = p_i - 2\,\sigma\,\kappa, \tag{4.5}$$

$$\partial_n v_t + \partial_t v_n = 0, \tag{4.6}$$

where v_t and v_n are the tangential and normal component of the velocity, respectively. Equation (4.6) results from the absence of tangential stresses. The left hand side of Eq. (4.5) is the normal force exerted by the fluid. It is the sum of the fluid pressure and viscous forces. The right hand side is the gas pressure which is reduced by the **capillary pressure** $2\,\sigma\,\kappa$, where κ denotes the average curvature[3] and σ the surface tension.

4.1.2 Gas Solubility and Diffusion

Besides pure hydrodynamics, gas diffusion and gas supply are essential for the evolution of foam. Generally, there is a finite solubility of the blowing gas in the fluid.

[3]The average curvature is defined as the mean value of the two principal curvatures, see also Section 9.2.1.

The gas concentration in the fluid is governed by the **diffusion equation**:

$$\partial_t c + v_\alpha \partial_\alpha c - \partial_\alpha \left(D \partial_\alpha c \right) = Q, \tag{4.7}$$

with $c = c(\mathbf{x}, t)$: concentration field, D: gas diffusion constant and $Q = Q(\mathbf{x}, t)$: source term. The source term Q describes the decomposition of the blowing agent.

In order to solve Eq. (4.7), boundary conditions have to be formulated. Normally, Dirichlet and von Neumann boundary conditions are distinguished. Whereas the gradients of the concentration field in normal direction are prescribed in the latter case, the concentration field is specified for the former one. It depends on the physical problem which approach is reasonable. In the case of foam, Dirichlet boundary conditions, where the concentration is given at the gas–liquid interface, is appropriate since the equilibrium concentrations at the interface can be calculated from **Henry's law**:

$$c(\mathbf{x}, t)|_\Gamma \propto p_i. \tag{4.8}$$

Henry's Law describes the dynamic equilibrium situation between a gas and a fluid saturated with this gas. The equilibrium gas concentration in the fluid is proportional to the partial pressure. Modifications to Henry's law are necessary if the state of the gas atoms in the gas is different from that in the fluid. For example, hydrogen is dissolved as H-atoms in a fluid whereas hydrogen gas consists of H_2-molecules. In this case, Eq. (4.8) transforms to the so-called **Sieverts' law**:

$$c(\mathbf{x}, t)|_\Gamma = S \cdot \sqrt{p_i}, \tag{4.9}$$

with Sieverts' constant S.

There are two basic subjects which have to be considered. The first issue is related to the solubility of the blowing gas which is nearly always hydrogen for metals. How much hydrogen gets dissolved in the liquid metal [237], especially at high pressures? Is there a danger of hampering foam formation during IFM due to the solubility of the blowing gas?

The second question refers to the diffusion velocity of the blowing gas. What are the typical diffusion distances during IFM? Have the blowing agent particles to be treated as isolated sources or is it legitimate to consider them as homogeneous background source?

4.1.2.1 Hydrogen Solubility

In order to elucidate the amount of dissolved hydrogen, the solubility is given in cm^3 hydrogen gas at normal conditions dissolved in a given mass of liquid metal. The solubility of hydrogen in pure aluminum is given by [175]:

$$c = \begin{cases} 5.84 \, \frac{\mathrm{cm}^3}{\mathrm{g}} \cdot \exp\left(-\frac{6357\,\mathrm{K}}{T}\right) \sqrt{\frac{p}{\mathrm{bar}}} & \text{liquid Al} \\ 0.25 \, \frac{\mathrm{cm}^3}{\mathrm{g}} \cdot \exp\left(-\frac{5941\,\mathrm{K}}{T}\right) \sqrt{\frac{p}{\mathrm{bar}}} & \text{solid Al.} \end{cases} \tag{4.10}$$

The solubility change at the phase transition at 660°C and $p = 1$ bar is $6 \cdot 10^{-3}$ cm^3/g.

The situation is different for magnesium. The solubility of hydrogen in pure magnesium is two orders of magnitude higher than that of aluminum and given by [194][4]:

$$c = \begin{cases} 9.31\,\frac{\mathrm{cm}^3}{\mathrm{g}} \cdot \exp\left(-\frac{2862\,\mathrm{K}}{T}\right)\sqrt{\frac{p}{\mathrm{bar}}} & \text{liquid Mg} \\ 2.07\,\frac{\mathrm{cm}^3}{\mathrm{g}} \cdot \exp\left(-\frac{1843\,\mathrm{K}}{T}\right)\sqrt{\frac{p}{\mathrm{bar}}} & \text{solid Mg.} \end{cases} \tag{4.11}$$

The solubility change at the phase transition at 650°C and $p = 1$ bar is 0.14 cm^3/g.

Figure 4.3 shows the hydrogen solubility for pure aluminum and magnesium as a function of the temperature.[5]

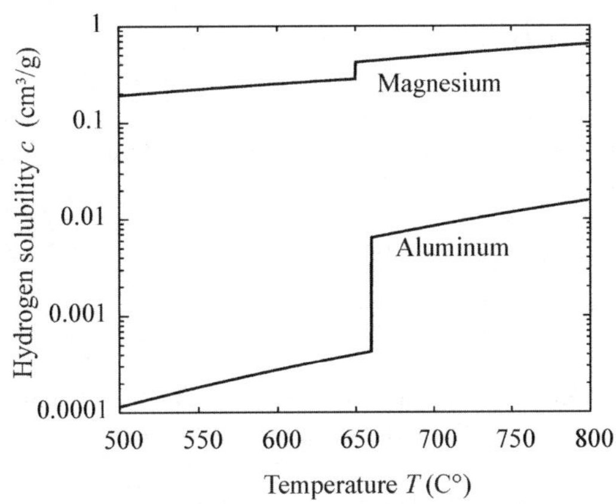

Figure 4.3: Hydrogen solubility for pure aluminum and magnesium at 1 bar, Eq. (4.10) and Eq. (4.11).

The hydrogen solubility in aluminum is not very high compared to the amount of hydrogen which is added during IFM[6] even if a pressure of 100 bar is assumed. Thus, for aluminum the efficiency of the blowing agent is not expected to be drastically reduced by the fact that part of the hydrogen gets dissolved. The situation is different if magnesium is considered. In this case, the solubility is two orders of

[4]See also [205].

[5]The hydrogen solubility depends on the particular alloy composition. Nevertheless, the addition of alloying elements does not change the solubility very much [194].

[6]As a rule of thumb, 1 wt% MgH_2 is added during IFM. This is about 25 cm^3/g under normal conditions.

magnitude higher. Nevertheless, the solubility is still small ($\approx$ 6 % for 1 wt% MgH_2) compared to the total amount of added gas even for foaming pressures of 100 bar.

4.1.2.2 Hydrogen Diffusion Length

A measure for the relevant length scale where mutual interaction of the different blowing agent particles can be expected is the so-called **diffusion length** δ^{diff} given by[7]

$$\delta^{\text{diff}} = \sqrt{4\,D\,t}, \tag{4.12}$$

where D denotes the diffusion coefficient and t the characteristic time.

The diffusion length is a measure for the region of influence of a blowing agent particle. The concentration field of each particle is autonomous if the distance between the particles is much larger than δ^{diff}. Strong mutual influence has to be expected if δ^{diff} is much larger than the mean particle distance. The diffusion length is also a means to assess whether a global blowing gas equalization in the component is possible on the relevant time scale of the process.

The diffusion coefficient of hydrogen in aluminum has been determined by Eichenauer and Markopoulus [64]:

$$D = D_0 \cdot \exp\left(-\frac{H}{RT}\right) = \begin{cases} 3.8 \cdot 10^{-6}\,\frac{\text{m}^2}{\text{s}} \cdot \exp\left(-\frac{19.26\,\text{kJ/mol}}{RT}\right) & \text{liquid Al} \\ 1.1 \cdot 10^{-5}\,\frac{\text{m}^2}{\text{s}} \cdot \exp\left(-\frac{40.95\,\text{kJ/mol}}{RT}\right) & \text{solid Al,} \end{cases} \tag{4.13}$$

where D_0 is a constant, R is the gas constant, T is the temperature and H the activation enthalpy. From Eq. (4.13), the diffusion coefficient at 700 °C is $D(700\,C^\circ) = 3.51 \cdot 10^{-7}\ \text{m}^2/\text{s}$.

The diffusion coefficient for hydrogen in solid magnesium is given by Wenzl [234]:[8]

$$D = D_0 \cdot \exp\left(-\frac{H}{RT}\right) = 4. \cdot 10^{-6}\,\frac{\text{m}^2}{\text{s}} \cdot \exp\left(-\frac{40\,\text{kJ/mol}}{RT}\right) \qquad \text{solid Mg.} \tag{4.14}$$

The extent of the diffusion zone can be estimated by applying Eq. (4.12). For $t = 0.1$ s, the diffusion length of hydrogen in liquid aluminum is 374 μm. This length is small compared to the overall dimension of the casting (10^{-2}–10^{-1} m). Thus, global concentration equalization cannot be expected. Consequently, inhomogeneous blowing agent distributions in the melt will lead to inhomogeneous porosities! This

[7] The diffusion length is defined as the distance to a spatial and temporal point-like source $Q = Q_0 \delta(t-t')\delta^{(3)}(\mathbf{x}-\mathbf{x}')$ where the intensity has dropped to e^{-1} (δ and $\delta^{(3)}$ denote the delta function in one and three dimensions, respectively). It is deduced from the fundamental solution of the diffusion equation, Eq. (4.7): $c(\mathbf{x},t) = Q_0 \cdot (4\,\pi\,D\,(t-t'))^{-\frac{3}{2}} \cdot e^{-\frac{(\mathbf{x}-\mathbf{x}')^2}{4\,D\,(t-t')}}$.

[8] The diffusion coefficient of hydrogen in liquid magnesium is not available in literature.

finding is confirmed by experimental observations. Foamed areas adjoin directly to not foamed areas if the blowing agent distribution is inhomogeneous.

Gas concentration equalization between different blowing agent particles depends on their mean distance (in the region of some 100 µm) given by the blowing agent particle density . Concentration peaks at blowing agent particles have to be expected on the relevant time scale. It is thus evident that blowing agent particles are good candidates to act as bubble nucleation sites.

4.1.3 Nucleation

A crucial step during foaming by in situ gas generation is nucleation. Bubble nucleation is expected to start when the solubility of the blowing gas in the melt is surmounted. However, for the evolution of a bubble nucleus a high amount of surface energy is consumed. That is why nucleation is – in most cases – not in the homogeneous melt (**homogeneous nucleation**) but at catalytic sites like small particles, oxides, etc. This kind of nucleation is named **heterogeneous nucleation** [50]. Nucleation at nucleation sites is more likely since the energy necessary for the development of the bubble interface is completely or partly compensated by the particle–liquid interface which is destroyed [41]. The fact, that nucleation is normally heterogeneous makes analytical descriptions difficult and information about nucleation has to be extracted from experiments.

During IFM, the blowing agent is added in form of a huge number of MgH_2 particles. These particles with a size of about 10–100 µm have a mean distance of some hundred micrometers and represent ideal nucleation sites, see Fig. 3.6. In addition, they are the source for the blowing gas. The estimate of the diffusion length in Sect. 4.1.2 shows that a balance of the local gas concentration is not possible during IFM. Consequently, the gas concentration is expected to develop local maximums at the blowing agent particles where the activation energy for nucleation is assumed to be small. These are very strong arguments for the expectation that nucleation takes preferably place at blowing agent particles. In fact, this assumption is confirmed by experimental observations, see Fig. 3.19.

If the nuclei density is too low, the resulting foam will automatically show a relatively large mean cell diameter at a given relative density. High nuclei densities are necessary to produce fine cell structures. Nevertheless, a high nuclei density alone does not necessarily lead to a fine foam structure, see Chapter 5.

4.2 Bubble Dynamics

Foams are agglomerations of interacting gas bubbles. The first step to understand the dynamics of foams is thus to have a closer look on the dynamics of isolated bubbles. There are various questions which have to be investigated. Is the development of bubbles possible on the very short time scale of IFM? Can viscous forces due to a strongly increased viscosity hamper bubble development? Is there a danger that the bubbles bubble up inside the casting due to gravity? What happens if two bubbles start to interact with each other?

The strategy of the following sections is to use analytical solutions for the behavior of isolated bubbles to deduce answers for the relevant time scale.

4.2.1 Single Bubble Expansion

Can it happen that bubble expansion is suppressed by viscous, capillary or inertia forces during IFM? From everyday experience we already know the answer. If an agitated bottle of mineral water[9] is opened gas bubbles develop very quickly. Thus, bubbles are expected to develop during a fraction of a second. Nevertheless, foam formation during IFM takes place while solidification takes place. Consequently, the viscosity of the semi-liquid metal can be strongly increased compared to the completely liquid state. In addition, the surface tension of metals is about a factor ten higher than that of water.

In order to estimate the relevant pressures we consider the expansion kinetics of an isolated bubble with radius R located in an infinitely extended incompressible liquid with viscosity ν and pressure p_0. The high symmetry of this problem allows the introduction of spherical coordinates thereby reducing the governing Navier-Stokes equations, Eqs. (4.1) and (4.2), to the so-called **Rayleigh equation** [197,201], which describes the temporal evolution of the bubble radius R:[10]

$$\underbrace{\rho R\ddot{R} + \frac{3}{2}\rho \dot{R}^2}_{\text{Inertia pressure}} + \underbrace{4\rho\nu\frac{\dot{R}}{R}}_{\text{Viscous pressure}} + \underbrace{\frac{2\sigma}{R}}_{\text{Capillary pressure}} = \underbrace{p_i - p_0}_{\text{Bubble overpressure}} , \quad (4.15)$$

where p_i denotes the bubble pressure and σ the surface tension.

The Rayleigh equation expresses equilibrium between inertia, viscous and capillary forces, which impede bubble expansion, and the driving force, i. e. the overpressure inside the bubble. The central question is, whether inertia, viscous or capillary forces play a dominant role or not, explicitly, is there a large overpressure in the bubble during growth or not. If yes, bubble expansion is delayed. If no, the forces on the left hand side of Eq. (4.15) are very small and equilibrium between bubble pressure, external pressure and capillary pressure can be assumed. The answer is obviously dependent on material and processing parameters and can not be universally given.

[9] The viscosity of water and liquid aluminum or magnesium is very similar.

[10] $\dot{R} \equiv \frac{dR}{dt}$ and $\ddot{R} \equiv \frac{d^2R}{dt^2}$.

In order to estimate the pressures, we assume that the temporal increase of the bubble volume V is proportional to its surface area:

$$\dot{V} = 4\,\pi\,R^2\,\dot{R} \qquad \text{with} \qquad \dot{R} = \text{const.} \tag{4.16}$$

With this assumption, the time t_f necessary to reach the finial radius R_f is given by

$$t_\mathrm{f} = \frac{R_\mathrm{f}}{\dot{R}}. \tag{4.17}$$

Figure 4.4 shows the different contributions to the bubble overpressure during expansion for the growth velocity $\dot{R} = 2.5$ mm/s.[11]

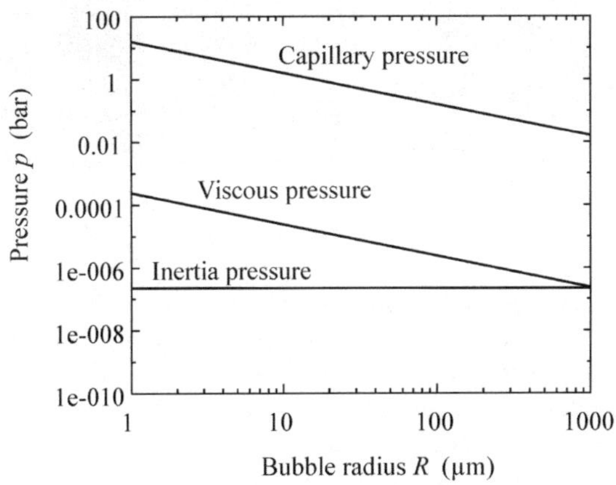

Figure 4.4: Pressures during the expansion of a single bubble in liquid aluminum, Eq. (4.15) ($\dot{R} = 2.5$ mm/s, $\nu = 1$ mm^2/s, $\sigma = 0.8$ N/m, $\rho = 2.4$ g/cm^3).

The viscous and the inertia pressures are obviously not significant on the relevant time scale. The capillary pressure dominates the bubble overpressure even if the viscosity of the melt is assumed to be four orders of magnitude higher than that of the completely liquid metal melt.[12]

In summary, a delayed bubble expansion due to viscous or inertia pressures can be ruled out during IFM even if the viscosity is strongly increased due to solidification. Thus, bubble expansion will follow the gas pressure inside the bubble without noticeable delay.

[11] This value is equivalent to the development of a bubble with radius 500 μm during 0.2 s.

[12] Liquid metal can achieve such high viscosities in the semi-liquid state, see Fig. 4.11. Whether this state in fact delays the expansion of very small bubbles is unclear since the bubble size is then in the same region or even much smaller than the solid particles in the melt. That is, the hydrodynamical description of this two-phase state in the continuum picture becomes invalid. See also Chap. 6.

4.2.2 Bubble Exposed to Gravity

Gas bubbles bubble up in a liquid very quickly. This effect might lead to a strong density gradient during IFM, see Fig. 4.5.

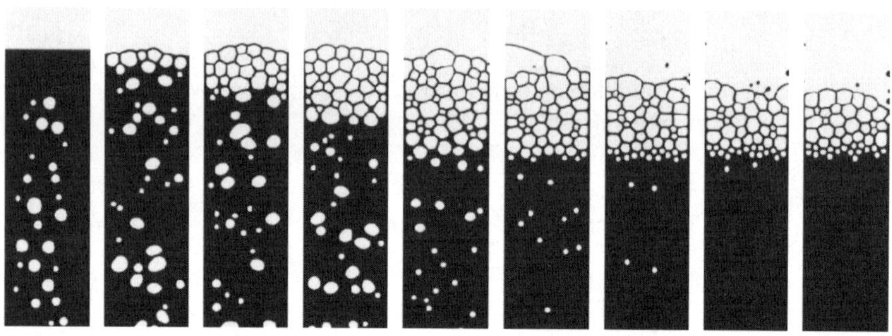

Figure 4.5: Bubbles exposed to gravity. Illustration of the formation of a foam head. (2D-LBM simulation)

In the following, a single gas bubble in an infinite fluid is considered. The bubble rises due to gravity and eventually reaches a stationary velocity v where the drag force balances buoyancy. If the deformation of the bubble is negligible, the stationary rising velocity v follows from force balance between the dragging force given by **Stokes' law** and buoyancy [168],

$$v = \frac{gR^2}{3\nu}, \tag{4.18}$$

where R is the bubble radius, ν the kinematic viscosity and g the gravity constant. The quadratic dependence of the rising velocity on the bubble radius expresses that small bubbles are overtaken by larger ones, see small bubbles in Fig. 4.5.

If the bubble velocity is assumed to fulfill Eq. (4.18), the covered distance d is given by

$$d = \int_0^{t_\mathrm{f}} v(R(t))\,dt = \int_0^{t_\mathrm{f}} \frac{g}{3\nu} R^2\,dt = \frac{g}{9\nu}\dot{R}^2\,t_\mathrm{f}^3 = \frac{g}{9\nu}R_\mathrm{f}^2\,t_\mathrm{f}, \tag{4.19}$$

with $R = \dot{R}\,t$ and $\dot{R} = \text{const}$. Assuming $R_\mathrm{f} = 0.5$ mm, $\nu = 1\ \mathrm{mm}^2/\mathrm{s}$, $t_\mathrm{f} = 0.2$ s and $g = 9.81\ \mathrm{m/s}^2$, the covered distance is about $d = 54$ mm! This value is an upper limit due to several reasons. The viscosity increases during solidification, the bubbles are not exactly round, stationarity is assumed and last but not least, bubble-bubble interaction is not taken into account. Nevertheless, this simple estimate shows that gravity might have an influence on the density distribution during IFM. Up to now, there is no experimental evidence for the significance of gravity. Nevertheless, larger parts might be more prone to this effect.

4.2.3 Bubble–Bubble Interaction

The primary interaction between bubbles results from hydrodynamical forces produced by the velocity fields induced by bubble movement or expansion. As a result, bubbles get deformed which again induces capillary forces trying to restore a spherical bubble shape in order to minimize the surface energy.

4.2.3.1 Transient Cell Walls

The most simple situation of two expanding bubbles is depicted in Fig. 4.6. Due to symmetry, the velocity field has to vanish in between the bubbles. As a result, bubble growth is asymmetric and the bubbles become more and more deformed during expansion. The magnitude of this effect depends on the surface tension, the viscosity, the growth velocity and the size of the bubbles.

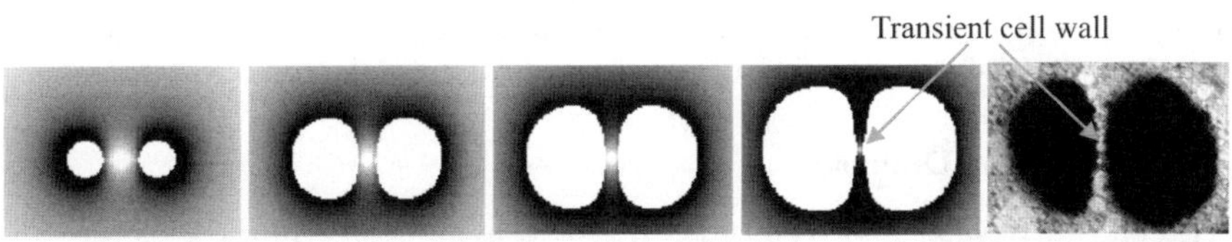

Figure 4.6: Bubble–bubble interaction and development of a transient cell wall. *Left*: 2D-LBM simulation (the gray scale indicates the velocity field; small velocities: bright, high velocities: dark). *Right*: Two bubbles frozen in magnesium (AZ91).

Figure 4.6 illustrates the development of transient cell walls. Transient cell walls are the result of rapid bubble expansion which brings the system far away from equilibrium (= round bubbles). Transient cell walls develop in a dynamic way and are intrinsically unstable, i. e. they vanish if the system has enough time to reach its equilibrium state. A characteristic gradient of the cell wall thickness indicates the transient state.[13] Whereas the cell walls are nearly flat with a curvature approaching zero, the Plateau borders show a curvature which is generally much higher than the mean curvature of the cell, see Fig. 4.7. The differences of the capillary pressure lead to over pressure in the walls and low pressure in the Plateau borders. The resulting sucking effect of this pressure difference leads to rapid cell wall thinning and eventually to cell wall rupture. This effect is called **drainage** and denotes the rearrangement of liquid under the influence of capillary forces.

The formation of transient cell walls is interesting with respect to foam formation. In principle, there should be a possibility to produce high porous, foam-like

[13]There are also **stable cell walls** where the capillary forces are balanced by a disjoining pressure whose range determines the cell wall thickness, see Sect. 4.3.

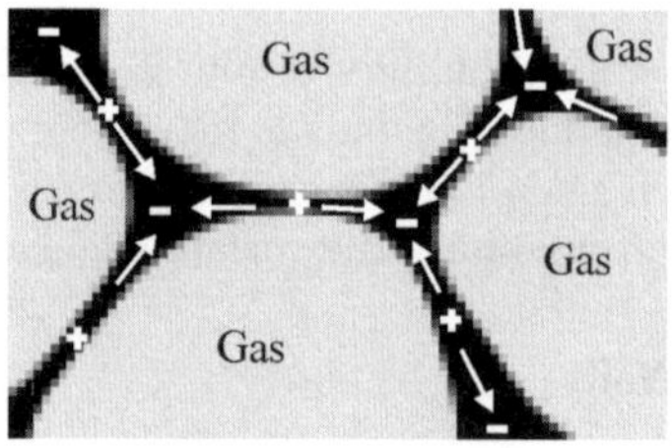

Figure 4.7: Drainage by capillary forces. The pressure difference between cell walls and Plateau borders leads to rapid cell wall thinning (over pressure is indicated by +, low pressure by −).

structures without stabilization by provoking the formation of transient cell walls. For this reason, it is instructive to estimate the life time of transient cell walls by estimating the cell wall drainage time until wall rupture takes place.

4.2.3.2 Cell Wall Drainage

Cell walls are instable and tend to be destroyed by drainage. In the following, the life time of cell walls is calculated with the help of a simple model for the cell wall. The continuous thinning of thin films, which represent the cell walls, between two parallel discs with rigid walls is regarded.[14] There is a capillary suction effect on the liquid from the cell wall to the Plateau borders. If R_{Pl} designates the curvature radius of the Plateau border the pressure difference Δp between Plateau border and cell wall reads

$$\Delta p = \frac{\sigma}{R_{\mathrm{Pl}}}. \tag{4.20}$$

The cell wall thinning velocity $\dot{d}$ is given by [21]:

$$\dot{d} = -\frac{2\,d^3\,\Delta p}{3\,\eta\,R^2} = -\frac{2\,d^3\,\sigma}{3\,\eta\,R^2\,R_{\mathrm{Pl}}}, \tag{4.21}$$

where R is the radius of the disc, η the viscosity of the liquid and Δp the pressure difference between the wall and the bulk, which is identified with the capillary pressure, Eq. (4.20). Integration of Eq. (4.21) gives the drainage time t with initial thickness d_{i} and final thickness d_{f}:

$$t = \frac{3\eta R^2\,R_{\mathrm{Pl}}}{4\sigma}\left[\frac{1}{d_{\mathrm{f}}^2} - \frac{1}{d_{\mathrm{i}}^2}\right] < \frac{3\eta R^2\,R_{\mathrm{Pl}}}{4\sigma}\frac{1}{d_{\mathrm{f}}^2}. \tag{4.22}$$

[14] The assumption of rigid walls gives us a lower bound for the draining velocity.

For $\sigma = 0.8$ N/m, $\eta = \nu\,\rho = 2.4$ g/(m s) [30], $R_{\mathrm{Pl}} = 100\ \mu$m, $R = 250\ \mu$m, and $d_{\mathrm{f}} = 10\ \mu$m[15] the thinning time can be estimated as

$$t < 10^{-4}\,\mathrm{s}. \tag{4.23}$$

This result shows that cell wall thinning is a very rapid process.[16] The thinning times are such short that the development of extensive, straight cell walls needs additional stabilization, which eventually stops thinning, see Sect. 4.3. On the other hand, if the viscosity is strongly increased, e. g. by three orders of magnitude – as it is the case during solidification – the wall thinning time approaches the same time scale as the whole foaming process.[17] Thus, the short process time in combination with the increased viscosity during solidification has the potential to support the formation of transient cell walls without further stabilization. This possibility of a dynamic way to stabilize a foam structure during evolution is discussed in detail in Chap. 5.

4.3 Stabilization Mechanisms

Decay is inherent to all compounded things.
Siddhartha Gautama (Buddha)

Pure liquids do not foam! A necessary prerequisite for the development of foam is the presence of a foaming agent. Generally, foaming agents can be specific adsorbed cations or anions from inorganic salts, polymers, particles, etc., which can often cause foaming at extremely low concentrations [199].

From a thermodynamical point of view, foams represent unstable states due to the high interfacial free energy and are therefore prone to decay. Two different classes of foams with respect to kinetics and the underlying mechanisms are commonly distinguished [199]:

1. **Unstable or transient foams** with lifetimes of seconds.
   In this case, surfactants[18] retard drainage but only weakly prevent film rupture. The foamability is thought to result from the Gibbs-Marangoni effect (Sect. 4.3.1) where a cell wall is stabilized during thinning due to material flow towards the weakened region because of the local increase in surface tension. The flow is the response of a surface tension gradient. Due to viscous drag

[15] This value is reasonable since thinner cell walls are not observed in integral foam [149].

[16] The more sophisticated approach in [34] leads to a similar result.

[17] The foaming time of IFM is smaller than one second whereas it is in the range of some minutes for all other known foaming methods.

[18] A surfactant is a substance that is capable to reduce the surface tension of a liquid in which it is dissolved.

the flow can carry an appreciable amount of underlying liquid along with it so that it restores the thickness.

2. **Meta stable or permanent foams** with lifetimes of hours until days. In these systems, the cell wall thinning times are relatively short compared to the lifetime. The stability is controlled by the balance of interfacial forces. These forces equilibrate after drainage has been completed. In the absence of external disturbances, i. e. evaporation, vibrations, temperature gradients or diffusion of gas, foam stability is almost infinite.

From the theoretical point of view, there is no universal theory for foam stability. Originally it was assumed that foam stability is due to adsorbed surfactants controlling the mechanical-dynamical properties of the surface layer. The first approaches to explain foam stability based on this idea were by Gibbs and Marangoni who introduced theories for surface viscosity and surface elasticity. Nevertheless, these early theories were not able to explain the high stability of thin films. In order to close this gap, the concept of a static disjoining pressure produced by intermolecular forces was introduced by Russian researchers in the middle of the 20th century.

We start this section with a general review of foam stabilization without special reference to metal foam. Dynamic and static stabilization are to be distinguished. Dynamic stabilization is produced by forces which develop from kinetics, e. g. viscous drag, whereas static stabilization comprises all mechanisms which generate static forces, e. g. electrostatic repulsion. Special emphasis is put on the role of solid particles with repect to foam stabilization. In addition, the state-of-the-art of metal foam stabilization is discussed.

4.3.1 Stabilization by Dynamic Forces

Foams are thermodynamically instable and prone to decay which is governed by drainage, cell wall thinning and subsequent cell wall rupture. In this section, mechanisms which slow down foam drainage kinetics and delay cell wall thinning and rupture in a dynamic way are discussed. These mechanisms usually act as resistance to thinning and rupture but do not contribute to the stabilization of a standing foam.

4.3.1.1 Dynamic Stabilization by Bulk Effects

As a general rule, the drainage rate of foam can be decreased by increasing the bulk viscosity of the liquid from which the foam is prepared, see Eq. (4.22). There are different methods available to increase the viscosity, e. g. addition of a solute, addition of a gel network forming substance or foaming in the semi-liquid state.

4.3.1.2 Dynamic Stabilization by Surfactants

Surfactants absorbed at the interface controll the mechanical-dynamical properties of the surface layer characterized by the **surface elasticity** and the **surface viscosity**. Surface viscosity reflects the relaxation speed of an imposed external stress whereas the surface elasticity is a measure of the energy stored in the surface layer as a result of the stress.

One effect of surfactants is to slow down cell wall thinning by increasing the surface viscosity [169] and the surface elasticity.[19] High concentrations of surfactants, polymers or particles in the surface cause high adhesive or cohesive bonding which increase both, the surface viscosity and the surface elasticity, see Fig. 4.8.

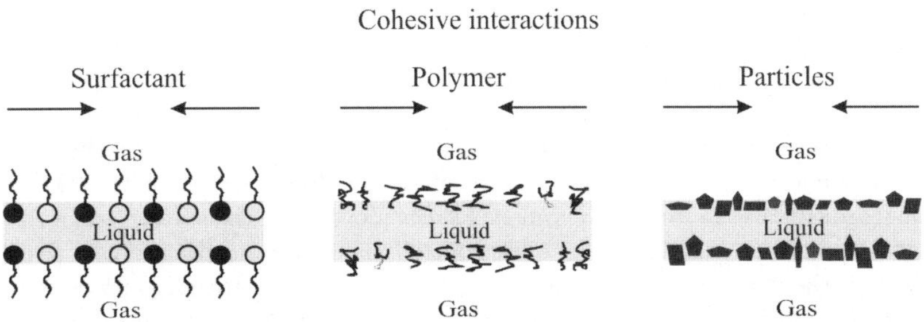

Figure 4.8: Different kinds of surfactants at the liquid-gas interface. *Left*: mixed surfactant system; middle: polymers; *right*: particles (after [199]).

The effect of the surface viscosity on the draining velocity of a cell wall can be explained as follows [169]. For high surface viscosities, the two adsorption layers behave nearly as rigid membranes and are not able to follow the vertical movement of the fluid. In this case, drainage follows the rules of liquid flow between fixed walls. If the surface viscosity is low, the absorption layer is mobile and able to follow the movement of the liquid and the resistance against drainage is even much smaller.

Starting point of cell wall rupture is a local thinning of the film which must be restored by an elastic force to stabilize the film. Theories describing such restoring effects were developed a long time ago by Marangoni and Gibbs (see references in [199]). The **Gibbs effect** describes the fact that a local thinning of the film will also lead to a local thinning of the surfactant density. As a result, a gradient in surface tension is induced creating a force which causes a material current in such a

[19]In fact, early experiments seemed to demonstrate a strong correlation between the dynamic stability of foam films and the surface viscosity. These early experiments led to the general concept that foaming and viscosity are always strongly interrelated [199].

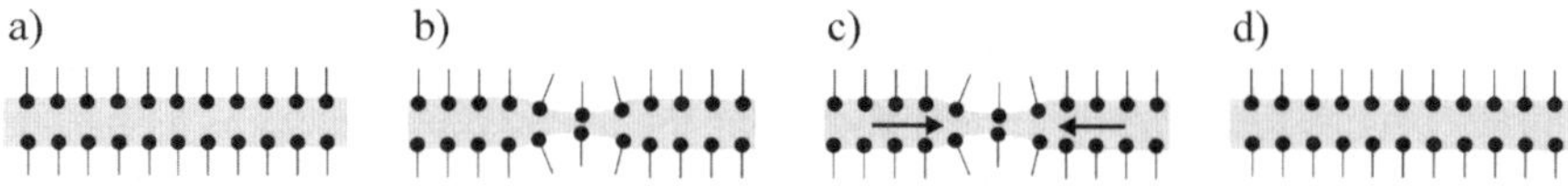

Figure 4.9: Gibbs-Marangoni mechanism of dynamic foam stabilisation. a) Original state of the thin film, b) Higher local surface tension due to local thinning, c) Gradient in surface tension pulls back surfactant molecules into thinned section, d) Surface film repaired by surface transport mechanisms.

way that the disturbance is balanced, see Fig. 4.9. If the surfactant concentration is too low, the induced force is also low and the disturbance can not be balanced. At very high surfactant concentrations the local thinning does not induce a gradient in surfactant concentration because there are surfactant molecules in the bulk which can immediately diffuse to the interface. As a result, there is a maximum in foam stability as a function of the surfactant concentration.

Besides the forces induced by the gradient of the surface tension there is an additional effect named as **Marangoni effect**. After local thinning, the non-homogeneously distributed surfactant molecules move along the surface. As a result, a thin liquid film near the surface is transported due to viscous drag forces. Accordingly, near the surface there is a fluid current from regions with low surface tension (high concentration) to regions with high surface tension (low concentration).

It is important to emphasize that both effects, Gibbs and Marangoni, are not sufficient to explain the existence of a stable foams where the thickness of the cell walls is more or less constant. That is, both effects are important to describe the behavior of transient foams. They are not able to explain the stability of permanent foams. How permanent foams are stabilized is described in the next section.

4.3.2 Stabilization by Static Forces

Prerequisite for the development of a permanent foam is the presence of static stabilizing forces. These forces work against the sucking effect of the Plateau borders and represent a kind of negative pressure in the cell walls. The different sources of static forces are discussed in the following.

4.3.2.1 Disjoining Pressure

The equilibrium or static stability of thin films or cell walls is extremely important. Foam films drain gradually under the action of gravity or capillary forces and eventually approach a metastable equilibrium state. Due to the fact that the capillary forces always tend to reduce the film thickness there must be an additional force, named **disjoining pressure** Π, balancing the capillary forces. The disjoining pressure comprises all forces that arise when the film surfaces are close enough to interact with each other. If Π is positive (repulsive), film thinning is suppressed. Otherwise, the driving force for film thinning is increased and film thinning is accelerated. In general, Π is computed as a sum of three contributions [21],

$$\Pi = \Pi_{\text{el}} + \Pi_{\text{vdW}} + \Pi_{\text{st}}, \tag{4.24}$$

with

- Π_{el}: Repulsive electrostatic double layer repulsion on the two surfaces.
- Π_{vdW}: Short range attractive van der Waals interactions.
- Π_{st}: Steric or structural interaction caused by steric hindrance in oriented and packed layers.

The range of Π_{el} as well as Π_{vdW} is very small, 10^{-9} m or even smaller. As a result, a foam stabilized by electrostatic double layer repulsion develops very thin stable cell walls (< 1 nm) which are commonly referred to as black films. In contrast to Π_{el} and Π_{vdW}, which are well understood [21], the origin of the repulsive steric interaction Π_{st} is still a subject of discussion.

4.3.2.2 Stability Criteria

Rupture of a cell wall occurs when disturbances on the surface grow in an unimpeded fashion. The dependence of the disjoining pressure on the wall thickness d determines whether a surface disturbance is damped or undergoes catastrophic growth. If local thinning leads to an increase of the disjoining pressure the local disturbance is cured and rupture is prevented. Otherwise, the local disturbance is accelerated and leads to rupture.

Wall stability is guaranteed as long as $\partial\Pi/\partial d < 0$. For the growth of instabilities and subsequent wall rupture it is necessary that $\partial\Pi/\partial d > 0$. The behavior of the disjoining pressure depends on the nature and size of the different contributions. Figure 4.10 shows one typical example. For this case, wall rupture will occur for $d < d_{\text{crit}}$. Wall rupture is therefore only possible if the capillary pressure exceeds the maximum disjoining pressure Π_{max}. For Π as in Fig. 4.10, the stability criteria of a wall are [21]:

$$\begin{array}{lll} d > d_{\text{crit}}: & \partial\Pi(d)/\partial d < 0 & \text{stable} \\ d = d_{\text{crit}}: & \partial\Pi(d)/\partial d = 0 & \text{metastable} \\ d < d_{\text{crit}}: & \partial\Pi(d)/\partial d > 0 & \text{unstable} \end{array} \tag{4.25}$$

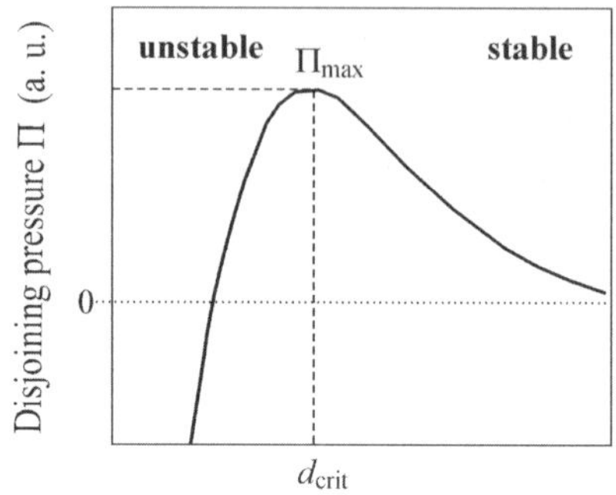

Figure 4.10: Disjoining pressure as a function of the cell wall thickness (Example, schematic). The cell wall is stabilized as long as the wall thickness exceeds d_{crit}. Smaller thicknesses lead immediately to cell wall rupture.

At the moment of rupture, the cell wall thickness is smaller than the cell wall thickness at the onset of instability. Cell wall rupture occurs more or less instantaneously at the moment when instability is reached. As long as

$$\frac{\sigma}{R_{Pl}} - \Pi(d) > 0, \tag{4.26}$$

liquid is sucked into the Plateau borders and the cell wall thickness continuously decreases. The **equilibrium cell wall thickness** is reached when the forces between the sucking effect of the Plateau borders and the disjoining pressure are balanced,

$$\frac{\sigma}{R_{Pl}} - \Pi(d) = 0 \qquad \Leftrightarrow \qquad \frac{\sigma}{R_{Pl}} = \Pi(d). \tag{4.27}$$

Consequently, the equilibrium thickness is not only determined by the disjoining pressure but also by the magnitude of the sucking effect of the Plateau borders, i. e. the surface tension. The magnitude of the disjoining pressure can be estimated from the mean Plateau curvature and the surface tension. High disjoining pressures lead to small curvature radii R_{Pl} of the Plateau borders. Very sharp Plateaus are thus a hint for the presence of a high disjoining pressure.

4.3.3 Stabilization by Particles

Most foams are based on a two phase system: a gas and a liquid containing a surfactant. The foams which we now focuse on comprise three phases: gas, liquid and solid particles. Dispersed particles can cause both, foam stabilization or destabilization [26].

4.3.3.1 Viscosity of Particle Suspensions

Dispersed particles increase the bulk viscosity of a liquid. Einstein found in 1906 that the viscosity ν of a particle suspension, whose particles can be treated as independent from each other, is a linear function of the relative particle volume content V_{rel} (= solid phase fraction) [168]:

$$\nu = \nu_0(1 + 2.5\,V_{\mathrm{rel}}) \qquad \text{for} \qquad V_{\mathrm{rel}} < 0.1, \tag{4.28}$$

where ν_0 denotes the viscosity of the particle free liquid. The **Einstein equation**, Eq. (4.28), indicates that the viscosity increase is quite moderate up to relative particle volumes of about 10%.

For higher relative particle volume contents, the particles are no more independent and there is no analytical description of the viscosity available. In addition, the particle size and shape are important in this regime. There are different approaches in literature which try to model the effect of higher relative particle volume contents. The most popular is the so-called **Krieger-Dougherty model** for spheres [145,236],

$$\nu = \nu_0 \left(1 - \frac{V_{\mathrm{rel}}}{V_{\mathrm{rel}}^{\mathrm{crit}}}\right)^{-2.5\cdot V_{\mathrm{rel}}^{\mathrm{crit}}}. \tag{4.29}$$

The critical relative particle volume content is denoted by $V_{\mathrm{rel}}^{\mathrm{crit}}$ and designates the volume content where flow is completely hampered by network formation. Various modifications of the Krieger-Dougherty model are available in literature [139]. Chen and Fan [39,68] derive for the viscosity of semi-liquid metal slurries,

$$\nu = \nu_0 \left(1 - \frac{V_{\mathrm{rel}}}{V_{\mathrm{rel}}^{\mathrm{crit}}}\right)^{-2.5}. \tag{4.30}$$

Figure 4.11 shows the relative viscosity as a function of the relative particle volume content for different $V_{\mathrm{rel}}^{\mathrm{crit}}$. For low V_{rel}, the Krieger-Dougherty model approaches the Einstein equation. The relative viscosity increase turns out to be very steep in a quite narrow range of V_{rel}. In addition, the viscosity increase is quite small up to particle volume contents of 30 %.

4.3.3.2 Bridging Mechanism

The increased viscosity slows down drainage and decay but does not contribute to foam stabilization. A contribution to stability is generated by particles due to their ability to bridge cell walls. This effect starts acting when the cell wall thickness is about equal to the particle diameter. The effect of the particles depends on their contact angle Θ with the fluid and their shape. The situation for spherical particles is depicted in Fig. 4.12.

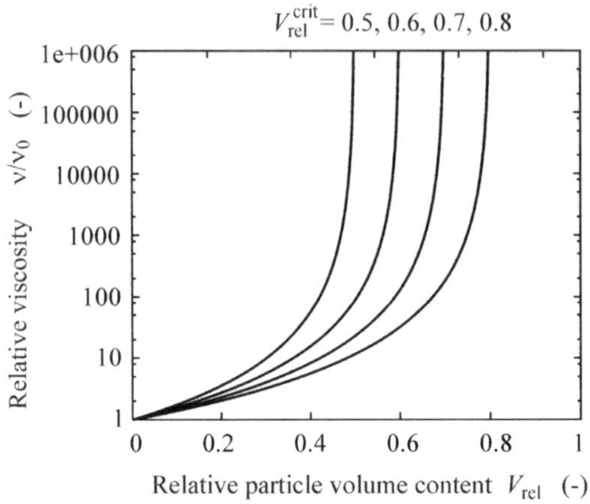

Figure 4.11: Viscosity of semi-liquid metal slurries, Eq. (4.30). Relative viscosity ν/ν_0 as a function of relative particle volume content V_{rel}.

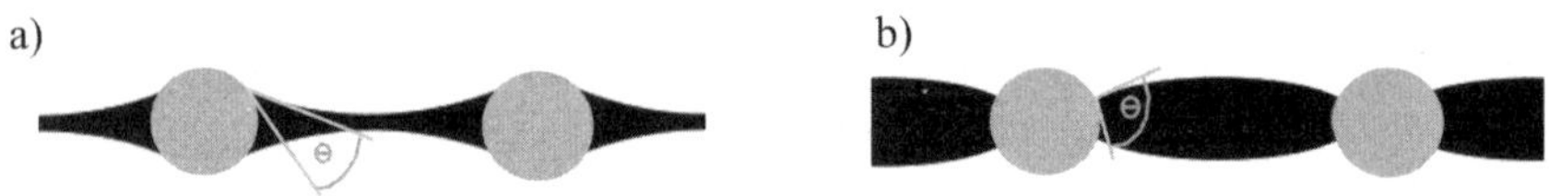

Figure 4.12: Influence of the contact angle on the bridging effect of particles. (**a**) Stabilization for $\Theta \leq 90°$. (**b**) Destabilization for $\Theta > 90°$ (schematic diagram after [124]).

Dippenaar [55] and Frye et al. [75] were the first who investigated the effect of the contact angle and the particle shape on the stabilization potential in a systematic way. Spherical particles with $\Theta \geq 90°$ stabilize films whereas they accelerate decay for $\Theta > 90°$. The reason for stabilization is the capillary pressure which sucks liquid to the particle. Destabilization results from the convex surface curvature which expels liquid from the particle. Non-spherical particles with special shapes are able to lead to film rupture even for $\Theta \leq 90°$. Thus, the wetting behavior in combination with the particle shape is essential to understand the action of particles.

4.3.3.3 Stratification

Foaming mechanisms in surfactant free particle suspensions have been comprehensively investigated by Wasan et al. [24,46,47,48,191,211,212]. For their experiments aqueous suspensions of nanosized hydrophilic silica particles were used. The foam is stabilized as the particles form a layered structure in the confined space of the foam lamella which provides a structural barrier against coalescence, see Fig. 4.13.

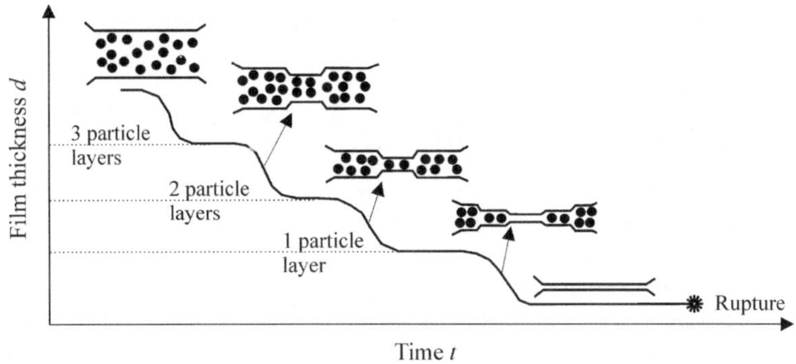

Figure 4.13: Stratification: Schematic of the temporal evolution of the film thickness (after [211]).

The existence of particle layering, which is referred to as stratification, is supported by the experimental observation of stepwise thickness transitions with the height of the steps approximately the size of the particles. The strength of the structural force and the number of step-transitions observed during film thinning is directly proportional to the particle volume content.

In order to understand the layering process, Chu et al. [48] use a numerical model where a model film consisting of equal size particles sandwiched between two rigid surfaces is considered. Two different kinds of pair-particle interactions are taken into account. A hard sphere model is used to model the strong steric-type repulsive interaction. The hard sphere model explains phenomena like particle layering, an oscillatory structural disjoining pressure and in-layer particles structures. The disjoining pressure generated by the particles turns out to be oscillatory with a period equal to about the effective particle diameter. Thus, the film thinning process is a manifestation of long-range structural forces, which show an oscillatory decay with thickness [26].

In addition to the repulsive interaction, an attractive force due to the so-called depletion phenomenon is taken into consideration. Depletion forces develop due to a difference in the osmotic pressure when the concentration of colloidal particles in the gap region between two particles differs from that in the bulk. This attractive force pushes two particles towards each other. The depletion force leads to new phenomena like particle condensation and the formation of voids inside particle layers which provide the explanation of black spot[20] formation.

[20]Black spots are regions in the film without any particles.

The fluid–gas interface is treated as a rigid wall in this model and the influence of the surface tension is not taken into account. In Chap. 6 we will see, that the deformation of the interface induces attractive forces between particles which are very similar to the depletion forces and which also lead to the phenomenon of particle condensation.

4.3.4 Metal Foam Stabilization

How metal foams[21] get stabilized is controversially discussed in the literature so far and a final, generally accepted theory is still lacking. This section is thought to summarize the main aspects of metal foam stabilization and especially to outline the physical idea which we have developed during several years of metal foam research [4, 45,121,146,147,149,158,160,161,163]. It became more and more evident that metal foam stabilization is based on one and the same mechanism, a barrier effect caused by particles confined in cell walls [146]. For different foaming methods, the nature of the particles acting as barriers is different. They can either be solid particles, fragile net-work particles or endogenous particles.[22] Their barrier effect always induces a disjoining pressure which is responsible for foam stabilization. This realization is the basis to explain stabilization of IFM which is discussed in detail in Chap. 6.

The first approaches to reveal the mechanisms of metal foam stabilization completely focused on the central role of the viscosity [12], bulk as well as the surface viscosity,[23] and the surface tension. A high viscosity retards cell wall thinning, see Eq. (4.22), and has thus a positive effect on the life time. The postulate is that metal foam stabilization and a high viscosity strongly correlate [196].

In fact, this idea seemed to be supported by several experimental observations. For example, during the Alporas process [181] calcium is added to an aluminum melt and stirred for several minutes until the viscosity reaches a specified value, which is by a factor of 3 to 4 higher than the initial one. Subsequently, TiH_2 is added as a blowing agent. The quality of the resultant metal foam is quite sensitive to the viscosity of the melt [176].

A further example are aluminum foams reinforced by SiC particles [7, 241], see Fig. 1.4. These foams are produced by gas injection into a metal matrix composite melt. The particle size ranges typically from 5 to 20 μm. If the particle volume content is too low, the injected gas bubbles up in the liquid and the bubbles burst at the top. The evolution of a stable foam is only possible if the volume content of SiC particles is high enough, i. e. if the viscosity is high enough.

[21] Here, metal foam comprises all cellular metals which develop in the liquid or semi-liquid state by the release or expansion of gas bubbles.

[22] Endogenous = originating or produced internally. The particles are not deliberately added but develop during solidification in the melt.

[23] The surface viscosity changes for e. g. by surface oxidation [7, 9].

An essential step to understand the underlying metal foam stabilization mechanisms is to have a closer look on their evolution behavior and the resulting cell structures. In the following, the main characteristics of metal foams are summarized:

- High stability
  Generally, metal foams show a high stability. On the one hand, foams produced via different production methods can be held in the liquid state for more than one hour without changing their internal structure [8, 9, 146, 181, 241]. On the other hand, real-time x-ray radioscopy reveals that cell wall rupture processes take place within less than 50 ms [15, 16, 214]. These properties are characteristic for permanent foams where a static interfacial force causes stability, see p. 92.

- Thick walls
  Typically, metal foams show thick cell walls of about 10^1–10^2 μm, see Fig. 1.4. A quite high volume fraction of material is located in the cell walls.

- Critical cell wall thickness
  During foaming of powder compacts (PM foam), cell coalescence processes take place throughout the whole expansion process [214]. Rupture takes place when the cell wall thickness falls below a critical value of about 50 μm. Thinner cell walls appear to be highly instable.

- Peculiar cell walls
  Typically, metal foams show strong, irregular thickness variations along cell walls [146], see Fig. 1.4. These stable thickness variations should be highly instable due to capillary forces.

- Non-ideal but stable cell structures
  The cell structures are far away from thermodynamical equilibrium if we assume that the surface energy of the structure has to be minimized, see Fig. 1.4. The mobility of the cell walls and Plateau borders seems to be strongly inhibited.

- Linear relationship between the cell diameter and the phase ratio
  The mean cell diameter D and the phase ratio ϕ are intimately correlated [53, 146, 147]:[24]

$$D \propto \phi = \frac{1}{\rho_{\mathrm{rel}}} - 1 \qquad \text{for} \qquad \phi > 1. \tag{4.31}$$

[24] In Chap. 5 it is shown that this relationship describes foam expansion where the mean material thickness is constant.

This relation expresses the fact that foam expansion is intimately related to cell coalescence processes.

It is important to realize that all theories which try to explain metal foam stabilization with the help of dynamical forces (i. e. the viscosity) are in contradiction with the most important experimental observation: **Metal foams show a high stability!**

The fact that metal foams are quite stable tells us that they are in a metastable state with a repulsive disjoining pressure responsible for stability. The range of the disjoining pressure can be deduced from the cell wall thickness and is about 10^1–10^2 μm. On the other hand, the thickness variations of the cell walls also indicate that the range can be locally quite different. The generally high material volume fraction in the cell walls is an indicator for a very high disjoining pressure which is able to suppress pronounced Plateau borders with a large radius of curvature, see Eq. (4.27). The presence of a high disjoining pressure becomes also apparent due to the presence of unfavorable cell arrangements. Foam structures far away from thermodynamic equilibrium demonstrate that the disjoining pressure is obviously so high that the surface tension is not sufficient to minimize the surface area.

Putting all together, we are looking for the origin of a long range force with variable range and a high magnitude. Forces with the range of μm can only be transmitted by body forces. Inspecting the different stabilization theories for foams, see Sect. 4.3, there is only one possible candidate which is able to explain metal foam stabilization: stabilization by particles. The stabilizing effect of the particles is primary not a surface effect but a bulk phenomenon. The surface tension prevents the particles from being expelled from the liquid, reduces their mobility and possibly forces them to arrange in a defined way in the cell wall. Since the particles are captured between interfaces they are able to transfer forces from one surface to the other by bridging the cell wall, see Fig. 4.12.

It is essential to state that the main effect of the particles is not to change the properties of the interface, i. e. surface tension or surface viscosity, but to represent mechanical barriers against cell wall thinning. The particles not only increase the bulk viscosity but also accumulate at the aluminum-gas interface if they are partially wet by the melt [80, 124]. Ip [124] was the first who realized that SiC particles can act as interface separators. Nevertheless, SiC particles are angular and only partially wet by liquid aluminum. Thus, a single layer of particles is expected to show a destabilizing effect [55, 75]. Experimentally, double layer and multi-layer arrangements are observed in straight cell walls. Single layer assemblies only occur as local indents. Kaptay [131,132,133,134] was the first who realized, that stabilization for this kind of particles is based on the development of 3D network structures which transfer forces from one interface to the other.

Besides deliberately added solid particles, there are also other structures which are able to operate as stabilizers. Necessary for the stabilizing effect is that mechanical

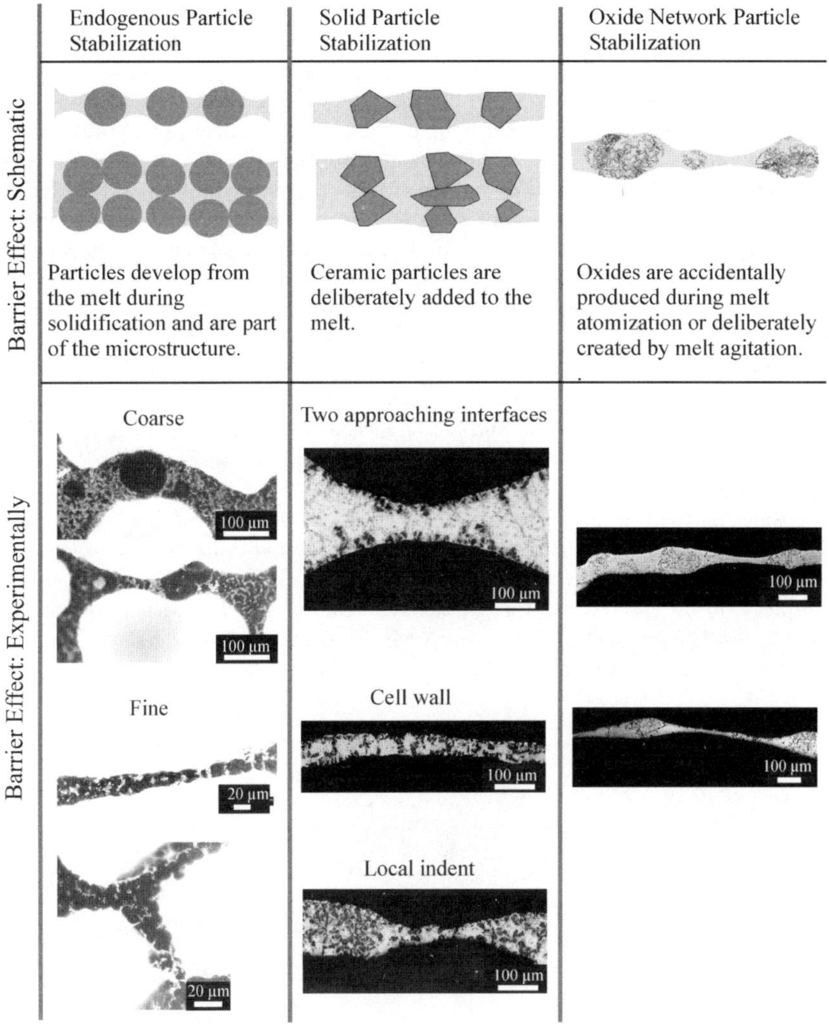

Figure 4.14: Metal foam stabilization by particle confinement. *Top*: Schematic diagrams illustrating the effect of particles on stability as the wall thickness approaches the particle diameter. *Bottom*: Light microscopies illustrating the barrier effect. *Left*: AZ91, LP-IFM, *Middle*: Al alloy/SiC, Cymat, *Right*: Al99.9, PM foam. The particles are represented by the dark phase [146].

forces are transmitted from one cell wall interface to the other. This task can not only be accomplished by solid particles but also by network structures, e. g. oxide fragments. The stabilization mechanism of foams produced by foaming of powder compacts (PM foam) was very long obscure. The only generally accepted fact was that the oxygen content of the underlying metal powder is essential for stabilization. The aluminum oxides were thus ascribed to increase the bulk viscosity or to decrease the surface tension. Both effects do not explain the observed stability of these foams. Furthermore, the oxygen mass content is so low, about 0.5–1.0 %, that a drastic increase of the viscosity can not be expected if the alumina is present as compact particles, see Eq. (4.28).[25]

Körner et al. [149, 157, 162] were the first who recognized that the oxides in the foamable precursor material form networks. The oxide films, which develop during atomization on the powder surface as well as in the powder particles, form loosely bounded, branched network structures. These network particles get captured in the cell walls during foam formation and act as mechanical barriers, see Fig. 1.4. Now, there is also evidence available that the Alporas foaming process is also based on oxides fragments as obstacles [8].

Considering today's knowledge of metal foam stabilization mechanisms there is only one conclusion which can be drawn: Metal foam stabilization is based on a barrier effect resulting from particle confinement, see Fig. 4.14. The particles can be either solid or loosely bound networks. Also the success of IFM is based on this principle, see Chap. 6. This is the reason why IFM foams show the same structure characteristics as metal foams produced by completely different processes.

[25] Microgravity experiments show that the influence of oxide particles on the bulk viscosity of the melt is of secondary importance while their main function is to prevent coalescence [227, 228].

5 Evolution Laws

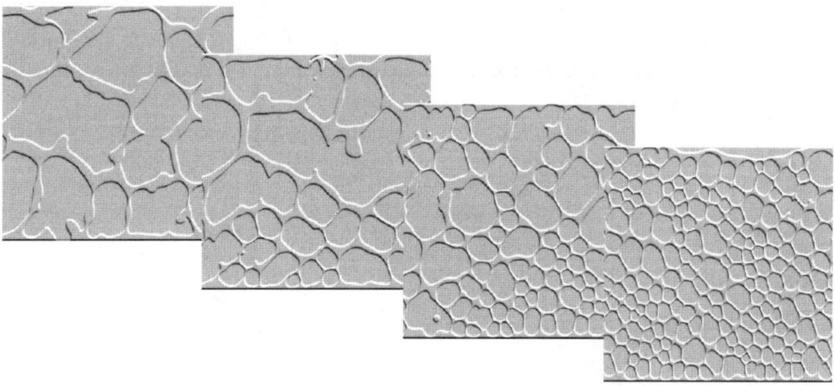

If it doesn't look easy it is that we have not tried hard enough yet.
Fred Astaire

During foaming, cell structures develop in a highly dynamic way. The residual cell configuration does not only depend on the material parameters like viscosity or surface tension, but also on the dynamics of the evolution process, the geometry and the boundary conditions. The structures deviate more and more from equilibrium when the velocity of the evolution process is increased. In sum, foam evolution is a very complex process!

Even so, the evolution of foam by in situ gas generation underlies simple general rules which we refer to as **evolution laws**. Different stabilization mechanisms lead to different evolution behavior. We apply both, numerical simulation[1] and analytical approaches in order to reveal the underlying principles.[2]

[1]The details of the numerical approach are described in Part III.

[2]Although foam evolution is caused by in situ gas generation, the resulting cellular structure is generally not very much influenced by the details of gas supply and diffusion. This is especially true for the fundamental evolution laws. Thus, aspects of the influence of gas diffusion on foam evolution laws are not discussed at this point. For details of the treatment of gas diffusion, see Appendix A.2.

5.1 Fundamentals

During foam evolution, there is an intimate correlation between the mean cell diameter D and the increasing ratio between gaseous and liquid phase ϕ. In many cases, D follows a power law,[3]

$$D = c \cdot \phi^n, \tag{5.1}$$

where c is a constant. We denote the exponent $n = n(\phi)$ as **evolution exponent**. Generally, n is characteristic for the underlying growth mechanism and determined by the way how foam stabilization – if at all – is realized.

Three stabilization categories characterized by different disjoining pressures Π are distinguished, see Fig. 5.1. Total stabilization, where cell coalescence processes

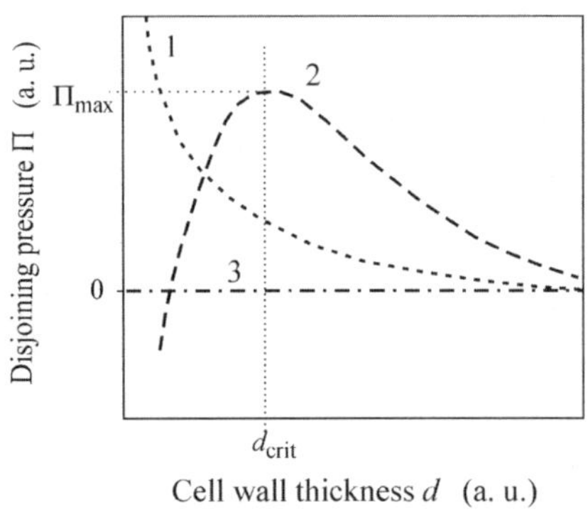

Figure 5.1: Stabilization categories. **1**. Total stabilization: $\Pi \longrightarrow \infty$ for $d \searrow 0$. **2**. Stabilization with upper bound: Π develops an absolute maximum with decreasing d. **3**. No stabilization: $\Pi = 0$.

are completely suppressed, is realized if the disjoining pressure increases infinitely during cell wall thinning. Things are different when Π has an upper bound. This regime only leads to cell wall stabilization as long as the critical cell wall thickness d_{crit} is not reached. If no disjoining pressure is present, cell walls are not stabilized, i. e. there is nothing which hampers cell coalescence.

In the following, foam evolution for the different stabilization categories is discussed and the evolution exponents are derived. We exploit this knowledge to deduce insight about the underlying mechanisms for the integral foam molding process, see Sect. 3.2.2.

[3]This result is different from that of Gibson and Ashby [85] who identified the relative density to be the relevant quantity to describe the mechanical properties of foam. ϕ and ρ_{rel} are connected, see Eq. (1.3).

Basis for the derivation of the evolution laws is the assumption that foam evolution can be characterized by two lengths scales, the **mean cell diameter** D and the **mean material thickness** δ.

The mean cell diameter D is calculated from the total gas volume V_G and the total number of bubbles n_B within the total volume V:

$$D = \sqrt[m]{\frac{2\,m\,V_G}{\pi\,n_B}}, \tag{5.2}$$

with $m = 2$ and $m = 3$ in two and three dimensions, respectively.[4]

The mean material thickness δ is a mean value of the thickness of the cell walls and the Plateau borders and is defined by

$$\rho_{rel} = \frac{(D+\delta)^m - D^m}{(D+\delta)^m}. \tag{5.3}$$

The phase ratio ϕ is thus given by

$$\phi = \frac{D^m}{(D+\delta)^m - D^m}. \tag{5.4}$$

5.1.1 Coalescence-Free Expansion

If coalescence events are not present or very rare, expansion is denoted as coalescence-free. This kind of expansion behavior appears more or less pronounced for all stabilization categories:

- Total stabilization: During all expansion states.
- Stabilization with upper bound: As long as the mean material thickness is clearly larger than d_{crit}.
- No stabilization: During the initial states of expansion.

Using Eq. (5.2), coalescence-free expansion of bubbles in two and three dimensions can be expressed as follows:

$$D = \sqrt[m]{\frac{2\,m\,V_G}{\pi\,n_0\,V_F}} = \sqrt[m]{\frac{2\,m}{\pi\,n_0}\,\phi}, \tag{5.5}$$

where the nuclei density n_0 is given by $n_0 = n_B/V_F$.

For coalescence-free expansion, the expansion exponent is determined to be $n = 1/2$ and $n = 1/3$ in two and three dimensions, respectively. In addition, it is

[4]In two dimensions, volumes have to be identified with areas.

particularly interesting to point at the dependence on the nuclei density which is the only dimensional factor in the absence of stabilization. Equation (5.5) applied to experimental data provides us with 3D information about the nuclei density n_0.

The higher the disjoining pressure the longer follows the mean cell diameter the analytical result of Eq. (5.5). Changes of the evolution mechanism are indicated by deviations from coalescence-free expansion. These deviations start rather abrupt and indicate the onset of cell coalescence processes.

5.2 Non-stabilized Evolution

A central question with respect to IFM is whether the evolution of foam structures in the absence of a disjoining pressure is possible. In this case, stabilization has to be entirely realized in a dynamic way, i. e. by a strongly increased viscosity. During IFM, we expect the high viscosity of the melt to generate such a dynamic stabilization effect due to parallel foaming and solidification.

Aim of this section is to derive the underlying evolution laws for the expansion of gas bubbles in a non-stabilized fluid ($\Pi = 0$). The influence of the viscosity and the capillary forces on the appearance of the cell structure and the evolution of the mean cell diameter is analyzed. Surprisingly, non-stabilized evolution follows simple general rules.

5.2.1 Cell Structure Evolution

Non-stabilized structure evolution is controlled by the interplay between viscous and capillary forces. In the following, the **capillary number** Ca is used to characterize the foam evolution process:

$$Ca = \frac{\eta v}{\sigma}, \tag{5.6}$$

with the dynamic viscosity η, the surface tension σ and the characteristic velocity v. We identify the characteristic velocity v as the expansion velocity of the foam. The capillary number represents the relative effect of viscous and capillary forces acting across a fluid–gas interface and determines the drainage time of cell walls, compare Eq. (4.21). Material rearrangement processes, which are necessary during foam expansion due to coalescence processes, are more and more hampered with increasing Ca.

The characteristic time for the decay of interfacial fluctuations is referred to as **capillary time** τ_{ca} [1],

$$\tau_{\mathrm{ca}} = \frac{\xi}{v_{\mathrm{ca}}}, \tag{5.7}$$

with the characteristic length scale ξ. The **capillary velocity** $v_{\rm ca}$ is defined by

$$v_{\rm ca} = \frac{\sigma}{\eta}. \tag{5.8}$$

As long as $v_{\rm ca}$ is considerably larger than the expansion velocity, i. e. $Ca \ll 1$, material rearrangement processes induced by bubble coalescence are still possible. Remnants of coalescence events are expected to become visible for $Ca > 1$.

The simulation of growing and coalescing gas bubbles in a completely liquid metal melt is quite difficult due to the extremely high surface tension in combination with the very low viscosity which results in capillary velocities of about 10^3 m/s, see Fig. 5.2.[5] These velocities are four or five orders of magnitude larger than the typical foam expansion velocities.[6] The viscous forces are very small compared to the restoring capillary forces. As a result, the duration of a coalescence event is much smaller than one millisecond for bubbles with radius of about 100 µm! Thus, coalescence of two bubbles is a highly dynamic process which impacts a larger neighborhood and eventually leads to sudden cascades of bubble coalescences, i. e. the melt boils. In between such cascades, the bubbles assume very quickly a perfect round shape.

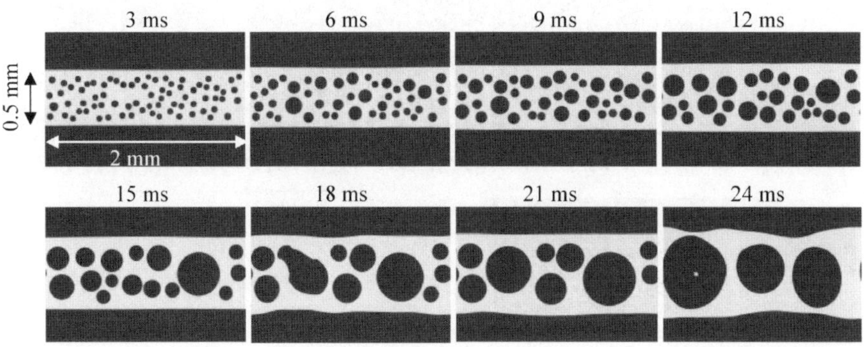

Figure 5.2: Bubble expansion and coalescence with $Ca = 0.82 \cdot 10^{-3}$ which is by a factor of 12.6 higher than the capillary number for liquid aluminum. (2D-LBM simulation, parameters: Table C.2)

[5]Liquid aluminum and magnesium show capillary velocities of $0.77 \cdot 10^3$ m/s and $0.45 \cdot 10^3$ m/s, respectively.

[6]The dynamics of gas bubble coalescence events in fully liquid metal melts is very fast and the reason why we use capillary numbers which are strongly increased compared to the real ones. Otherwise, we run into numerical instabilities.

From Fig. 5.2 it becomes clear that foaming of non-conditioned metal melts in the fully liquid state has no reasonable chance due to massive coarsening. However, during IFM the viscosity and thus Ca are expected to be strongly increased compared to that of the liquid metal since foaming takes place parallel to solidification. The influence of Ca on the resulting cell structure is depicted in Fig. 5.3. As ex-

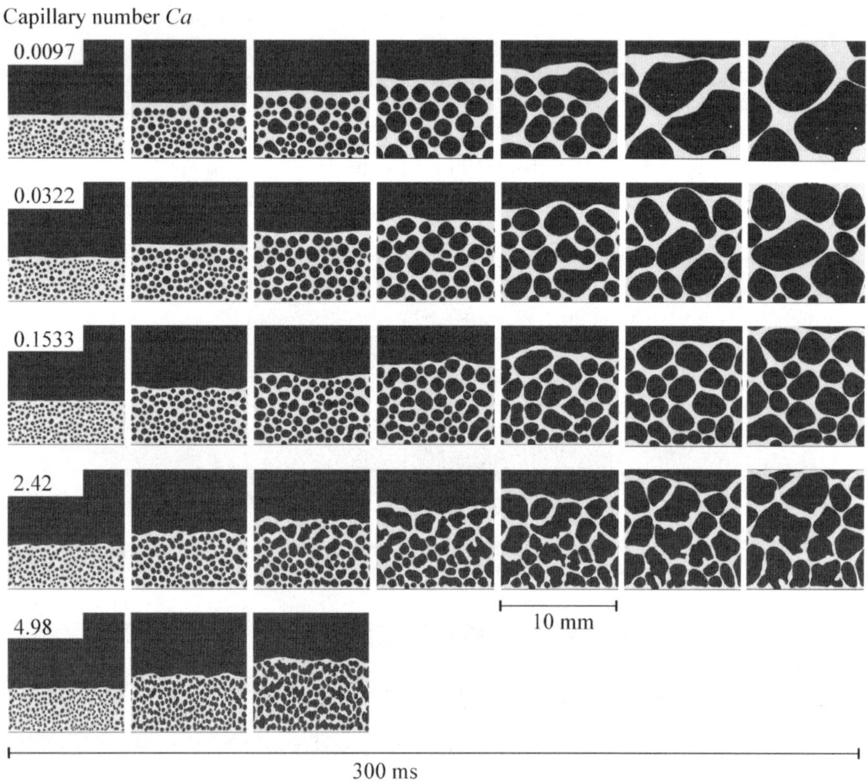

Figure 5.3: Influence of the capillary number Ca on the structure evolution in aluminum during 300 ms. Capillary forces control the expansion behavior for $Ca < 1$ whereas viscous forces dominate for $Ca > 1$. $Ca = 1.3 \cdot 10^{-4}$ for liquid aluminum. (2D-LBM simulation, parameters: Sim2 - Sim6, Table C.3)

pected, Ca has strong influence on the appearance of the cell structure. For $Ca < 1$, expansion is yet similar to boiling and coarsening is very fast. Cascade-like bubble coalescence events are more and more suppressed by increasing Ca and the resulting

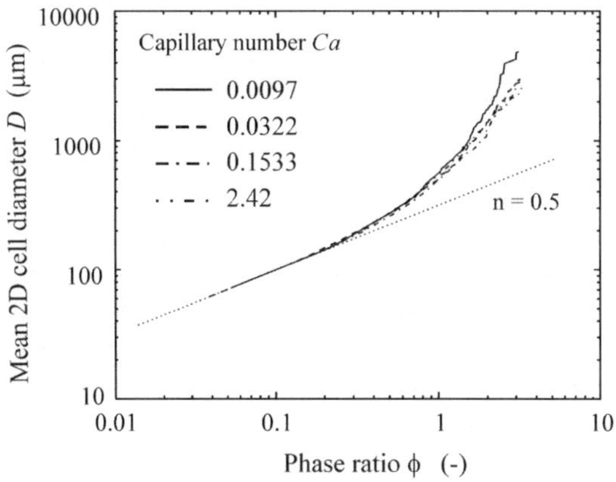

Figure 5.4: Mean cell diameter for various capillary numbers as a function of the phase ratio ϕ. (2D-LBM simulation, parameters: Sim2 - Sim5, Table C.3)

cell structure appears foam-like. A high Ca not only slows down cell wall thinning and subsequent cell coalescence but also delays spheroidization. A further increase of Ca eventually leads to completely disrupted structures since spheroidization is no longer possible.

Numerical simulation shows that high viscosities and low surface tensions, i. e. large capillary numbers, are favorable to decrease the drainage rate and to suppress cascade-like coalescence events. Nevertheless, a high Ca does not lead to a structure with polygonal cells with straight cell walls but rather to disrupted structures. Thus, small capillary numbers end up with severe coarsening whereas high capillary numbers result in completely disrupted structures.

In order to obtain a more quantitative description of the influence of Ca it is helpful to consider the mean cell diameter D as a function of the phase ratio ϕ, see Fig. 5.4. Coalescence events start to happen very early. The range of the phase ratio with the mean cell diameter following coalescence-free expansion ($n = 0.5$)is thus rather short. It is interesting to note that the deviation from the coalescence-free behavior starts more or less simultaneously for all capillary numbers. Obviously, Ca does not have strong influence on the onset of coalescence. Furthermore – this is the much more interesting observation – the influence of Ca is generally rather small, also for higher phase ratios! Only the data for the lowest Ca shows small deviations from the general behavior for large ϕs.[7] This insensitivity has not been expected but

[7]To what extent this deviation is a result of the finite system size in combination with cascade-like coarsening effects is unclear.

a clear decrease of the mean cell diameter with increasing Ca. Obviously, the idea to stabilize an expanding bubble system simply by increasing the viscosity does not work very well. Anyhow, the fact that the system behavior shows to be insensitive to Ca points to a simple general behavior.

5.2.2 Evolution Law

In non-stabilized bubble systems, the mean cell diameter grows very rapidly when higher phase ratios are approached. What happens? Given a fixed number of bubbles there is a maximum phase ratio $\phi_{\max}$ without bubble deformation. A further increase of ϕ would automatically lead to bubble deformation and subsequent bubble coalescence due to the absence of stabilization. Trying to approach $\phi_{\max}$ will thus lead to a divergence of the mean cell diameter.

What determines $\phi_{\max}$? The maximum phase ratio is not a constant but is dependent on the bubble diameter distribution. In principle, arbitrary high values of $\phi_{\max}$ may be realized with bubbles of suitable diameters. Generally, during foaming the situation is very similar to one with only one bubble diameter, the mean cell diameter. Thus, $\phi_{\max}$ is expected to be determined by the relative density of some sort of sphere packing.[8]

In order to obtain $D = D(\phi)$, an analytical expression for the mean material thickness $\delta = \delta(D, \phi)$ has to be derived. To find such an expression, a thought experiment is performed, see Fig. 5.5. Our basic hypothesis is scale invariance of the problem. It is assumed that the resulting structure should not depend on the overall length scale. If the length scale is now increased by a factor of k, the nuclei density n_0 will decrease by a factor of $1/k^m$. On the other hand, the mean cell diameter D will be increased by a factor of k. Due to the fact, that the phase ratio is assumed to be independent from the length scale, ϕ can be expressed as follows

$$\phi = f\left(D \cdot \sqrt[m]{n_0}\right), \tag{5.9}$$

with f denoting a continuously differentiable function, and $m = 2$ and $m = 3$ in two and three dimensions, respectively. Comparing Eq. (5.9) with Eq. (5.4)

$$\phi = \frac{D^m}{(D+\delta)^m - D^m} = \frac{1}{\left(1 + \frac{\delta}{D}\right)^m - 1} \tag{5.10}$$

leads to

$$\delta = D \cdot g\left(D \sqrt[m]{n_0}\right), \tag{5.11}$$

with g denoting a continuously differentiable function.

[8] $\phi_{\max} = 1.78$ for random sphere packings in 3D [125].

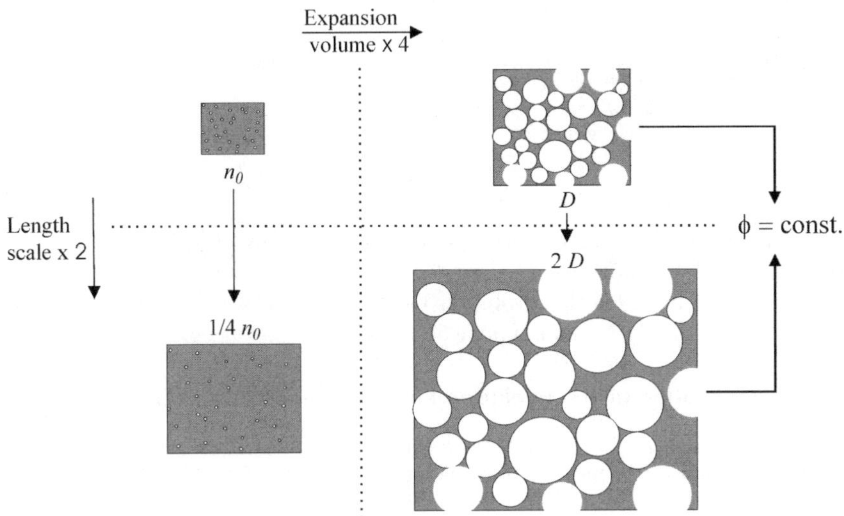

Figure 5.5: Scale invariance. Increasing the length scale by a factor of two will reduce the nuclei density by a factor of four and increase the mean cell diameter by a factor of two while keeping the phase ratio constant.

What do we know about the function g? The fact that the phase ratio should increase with increasing mean cell diameter restricts the power series expansion of g:

$$g\left(D\ \sqrt[m]{n_0}\right) = \sum_{i\leq 0} c_i \cdot \left(D\ \sqrt[m]{n_0}\right)^i, \tag{5.12}$$

with the constants c_i. The mean material thickness δ can now be expressed by the first two leading terms given by

$$\delta \approx c_{-1}\,\frac{1}{\sqrt[m]{n_0}} + c_0\,D. \tag{5.13}$$

The last expression for the mean material thickness δ expresses the fact that there are only two lengths scales available in the non-stabilized bubble system, $1/\sqrt[m]{n_0}$ and D.

Combining Eq. (5.13) with Eq. (5.10) provides us the **evolution law for non-stabilized bubbles**:

$$D = \frac{c_{-1}}{\sqrt[m]{n_0}}\,\frac{1}{\sqrt[m]{\frac{1+\phi}{\phi}} - 1 - c_0}. \tag{5.14}$$

For $\phi \gg 1$, D is given by the leading terms of Eq. (5.14)

$$D \approx \frac{m\, c_{-1}}{\sqrt[m]{n_0}} \frac{\phi}{1 - m\, c_0\, \phi} = \frac{c_{-1}}{\sqrt[m]{n_0}\; c_0} \frac{\phi}{\phi_{\max} - \phi} \qquad \text{with} \quad \phi_{\max} = \frac{1}{m\, c_0}. \tag{5.15}$$

Eq. (5.15) reveals a divergence at

$$\phi = \phi_{\max} = (m\, c_0)^{-1}. \tag{5.16}$$

The mean cell diameter is predicted to be proportional to $1/\sqrt[m]{n_0}$.[9] The representation of Eq. (5.15) in a power series of ϕ illustrates that an evolution coefficient n can not be defined for this evolution mode. The evolution exponent increases continuously for $\phi \nearrow \phi_{\max}$ and approaches infinity.

5.2.2.1 Verification of the Evolution Law for Non-stabilized Bubbles

In order to verify Eq. (5.14) numerical experiments are compared with the theoretical prediction. The mean cell diameter as a function of ϕ for different initial nuclei densities is represented in Fig. 5.6. Besides the numerical result (full lines), coalescence-free growth (doted lines) and the evolution law for non-stabilized bubbles (dashed lines) are depicted. After determination of the two free parameters, c_{-1} and c_0, the functional dependence for all three nuclei densities is reproduced! This is a very strong hint, that Eq. (5.14) covers the basic physical mechanisms during non-stabilized evolution.

An important issue is the influence of the initial nuclei density on the foam structure. Intuitively it is clear that a high nuclei density is favorable. But how high should it be? Is it possible to generate foams with a very fine structure and a very low relative density simply by increasing the nuclei density or is there an upper limit where a further increase of the nuclei density does not help to improve the cellular structure?

Due to the $1/\sqrt[m]{n_0}$-dependence of D in Eq. (5.14) it is instructive to normalize the numerical data by multiplying the mean cell diameter with $\sqrt[m]{n_0}$, see Fig. 5.7. In sum, eight different numerical experiments are depicted with the nuclei density varying from $6.25/\text{mm}^2$ to $39/\text{mm}^2$ and the capillary number from 0.0097 to 2.42. As predicted, all experiments lie more or less on the same master curve! Deviations from the general behavior probably result from the finite system size. Coalescence-free growth dominates for $D < 1/\sqrt[m]{n_0}$ and non-stabilized evolution is expected for $D > 1/\sqrt[m]{n_0}$.

For coalescence-free as well as non-stabilized evolution, $D \propto 1/\sqrt[m]{n_0}$. However, the divergence at $\phi_{\max}$ expresses the fact that increasing n_0 does not help to generate cellular structures with $\phi > \phi_{\max}$.

[9] Note that $1/\sqrt[m]{n_0}$ is the mean nuclei distance.

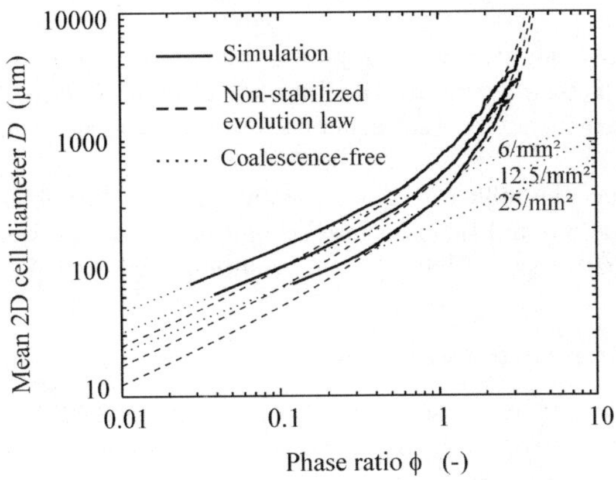

Figure 5.6: Verification of the non-stabilized evolution law. Numerical experiments for various nuclei densities (full lines: 2D-LBM simulation, parameters: Sim3, Table C.3) are compared with the non-stabilized evolution law (*dashed lines*: Eq. (5.14) with $c_{-1} = 0.55, c_0 = 0.105$ and $\phi_{\max} = 1/(2 \cdot c_0) = 4.76$).

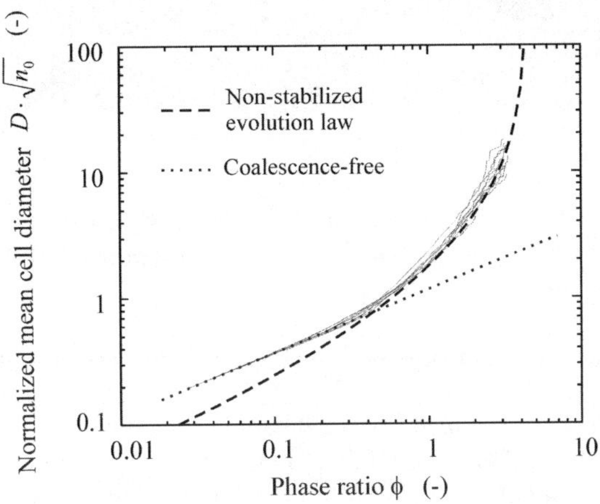

Figure 5.7: Master curve for non-stabilized evolution. Normalized mean cell diameter for 8 numerical experiments with the nuclei density ranging from $6.25/\text{mm}^2$ to $39/\text{mm}^2$ and the capillary number Ca from 0.0097 to 2.42. (Non-stabilized evolution: Eq. (5.14) with $c_{-1} = 0.55, c_0 = 0.105$ and $\phi_{\max} = 1/(2\ c_0) = 4.76$, 2D-LBM simulation: parameters: Sim2 - Sim6, Table C.3)

5.3 Stabilized Evolution

In the presence of a non-vanishing disjoining pressure foam evolution is denoted as stabilized. The disjoining pressure Π comprises all kind of acting forces which are not reliant on movement, i. e. the disjoining pressure does not vanish for standing foams.[10]

In the following, the influence of a disjoining pressure[11] on foam expansion, the resulting cell structure and the associated evolution law is investigated. Analogous to the non-stabilized case, stabilized foam evolution is governed by simple laws.

5.3.1 Cell Structure Evolution

Cell coalescence events are retarded or completely suppressed in the presence of a disjoining pressure. An example for the cell structure evolution in the presence of a bounded Π is depicted in Fig. 5.8.[12]

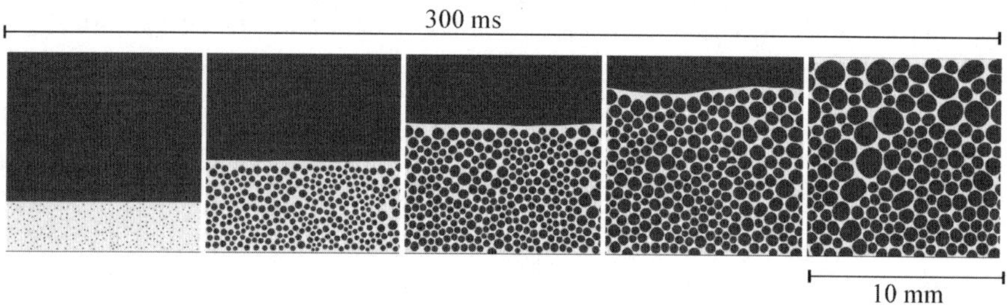

Figure 5.8: Cell structure evolution for stabilization with upper bound and $\frac{\Pi_{\text{max}}}{\sigma/\Delta x} = 0.334$. Cell coalescence is delayed but not completely suppressed. (2D-LBM simulation, parameters: Sim3, Table C.3)

The stabilizing effect increases with increasing Π for low as well as for high viscosities, see Fig. 5.9.

The high mobility of the melt for small capillary numbers leads to a quite round and uniform cell shape. For large capillary numbers, the appearance of the cell structure is much more angular and the cell diameter distribution broadens. A further

[10] The origin of a disjoining pressure for IFM is discussed in detail in Chap. 6.

[11] For the functional dependence of the disjoining pressure used for numerical simulation, see Sect. 7.4.

[12] The magnitude of stabilization is governed by $\Pi_{\text{max}}/(\sigma/\Delta x)$, compare Eq. (4.27). The equilibrium Plateau border radius R_{Pl} normalized to the lattice length scale Δx is given by $R_{\text{Pl}}/\Delta x = (\sigma/\Delta x)/\Pi_{\text{max}}$.

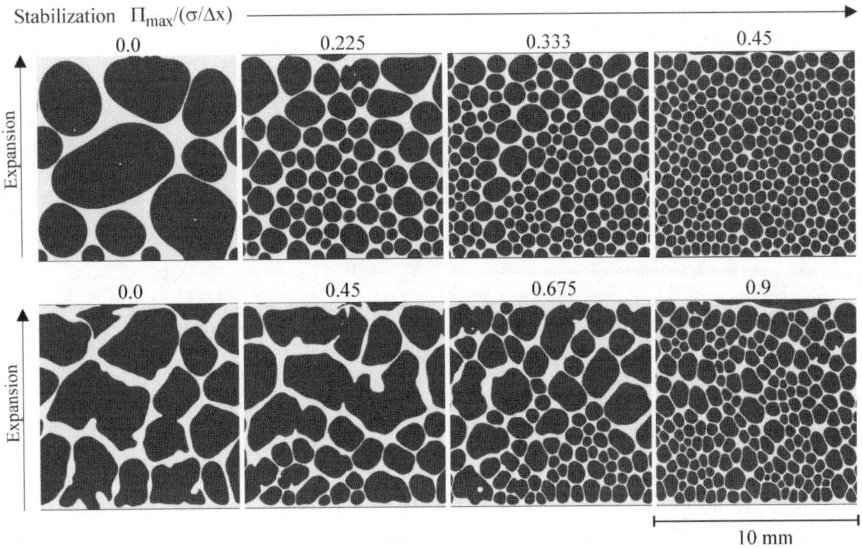

Figure 5.9: Cell structures after 300 ms for different disjoining pressures and capillary numbers. *Upper row:* $Ca = 0.0322$. *Lower row:* $Ca = 2.42$. $Ca = 1.3 \cdot 10^{-4}$ for fully liquid aluminum. (2D-LBM simulation, parameters: Sim3 + Sim5, Table C.3)

observation is that coarsening of the cell structure starts at the front of the foam where the movement and the strain rates are maximal. This effect demonstrates that the resulting structure is not only a function of the material parameters but also dependent on the velocity of foam expansion and the geometry of the surrounding cavity.

The evolution of the mean cell diameter as a function of the phase ratio is depicted in Fig. 5.10. For small Ca, the mobility of the melt is high enough to allow cell rearrangements due to the repulsive action of the disjoining pressure. The behavior is quite different if Ca is large. The low mobility of the melt acts to prevent rearrangement processes. As a result, cell coalescence processes start much earlier although stabilization is higher. The disjoining pressure suppresses cell wall rupture in an efficient way. However, if Ca is large, the stabilizing effect of the disjoining pressure is strongly reduced since it is frequently surmounted by local viscous forces which prevent necessary rearrangements of the cells. Thus, the dynamics of the evolution process has a strong effect on the cell structure. High expansion velocities or high viscosities lead to cell structures far away from equilibrium since rearrangement

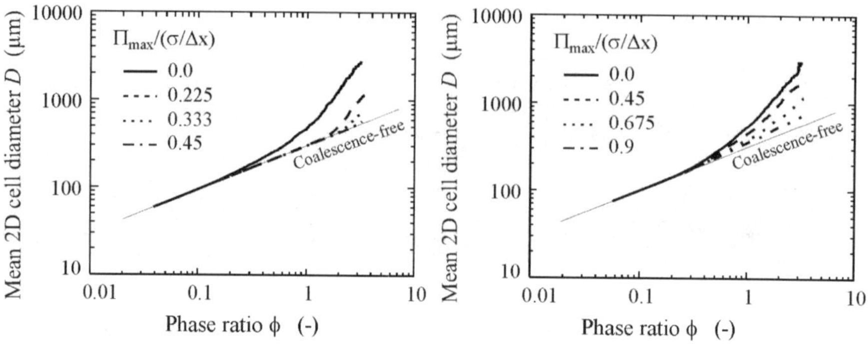

Figure 5.10: Mean cell diameter for the simulations in Fig. 5.9 as a function of the phase ratio. *Left*: $Ca = 0.0322$. *Right*: $Ca = 2.42$.

processes are no more possible. The intensity of the deviation from the equilibrium structure can be estimated from the appearance of the cells. Generally, the emergence of angular, stretched cells and remnants of cell rupture events are strong hints for non-equilibrium expansion, see Fig. 5.9. If the deviations from equilibrium are limited, the expansion is governed by a phenomenon which we refer to as **growth coalescence**, see Sect. 5.3.2.

5.3.2 Growth Coalescence

Growth coalescence designates the phenomenon that expansion of a cell structure is only possible on the expense of cell coalescence [146]. Consequently, the total cell number in a foam decreases with decreasing relative density. Necessary for the appearance of growth coalescence is the presence of a disjoining pressure with upper bound. On the one hand, the disjoining pressure stabilizes the cell structure and balances the sucking pressure from the Plateau borders, see Eq. (4.27). Thus, Π causes each individual expansion state to represent a state near equilibrium, i. e. if expansion stops the actual cell structure is maintained. On the other hand, cell coalescence events are not completely suppressed due to the limitation of Π. During expansion, the cell walls as well as the Plateau borders get depleted from liquid as long as the disjoining pressure increases. When the critical cell wall thickness is reached, cell coalescence takes place.

In order to reveal the underlying evolution law for growth coalescence it is necessary to investigate foam expansion up to very high phase ratios. Figure 5.11 shows the cell structure evolution for low and high stabilization where the final phase ratio is about 10. The Plateau borders, which are clearly visible in the case of low stabilization, nearly vanish when stabilization is high.

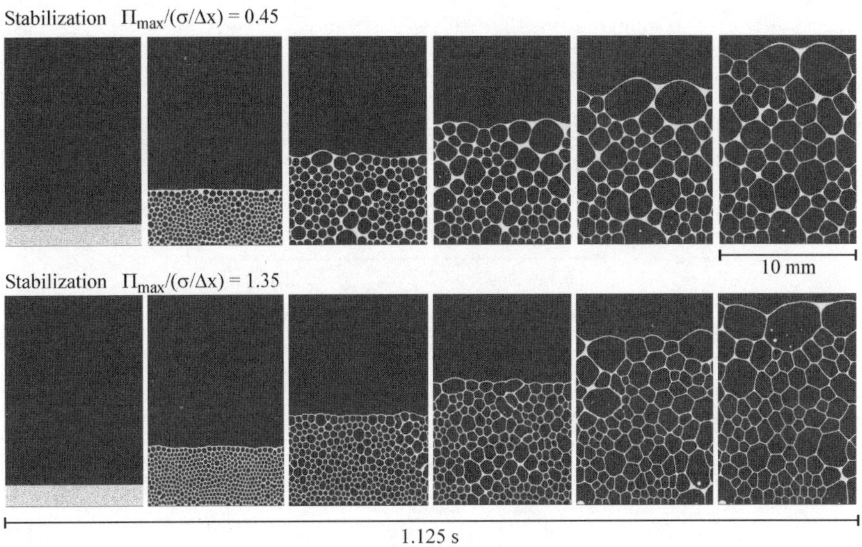

Figure 5.11: Growth coalescence. *Upper row:* Low stabilization. *Lower row:* High stabilization. (2D-LBM simulation, parameters: Sim7 + Sim8, Table C.4)

The evolution of the mean cell diameter for the two examples in Fig. 5.11 is depicted in Fig. 5.12. Obviously, the growth mechanism changes quite abruptly at a phase ratio of about one to two. For larger phase ratios, the mean cell diameter follows a relationship corresponding to a slope of one in the plot. Obviously, stabilized foam evolution seems to underly simple rules.

The first step to reveal the evolution law for stabilized growth is to find an expression for the mean material thickness δ. For non-stabilized evolution, compare Eq. (5.13), the mean nuclei distance $1/\sqrt[m]{n_0}$ and the mean cell diameter D are the only length scales available. For stabilized evolution, the range of the disjoining pressure provides an additional length scale.

The mean material thickness δ is assumed to be independent from the nuclei density and the cell diameter. The underlying assumption for growth coalescence is

$$\delta = \text{const.} = \delta(\Pi). \tag{5.17}$$

From this assumption and Eq. (5.4), the **evolution law for stabilized foams** is derived as

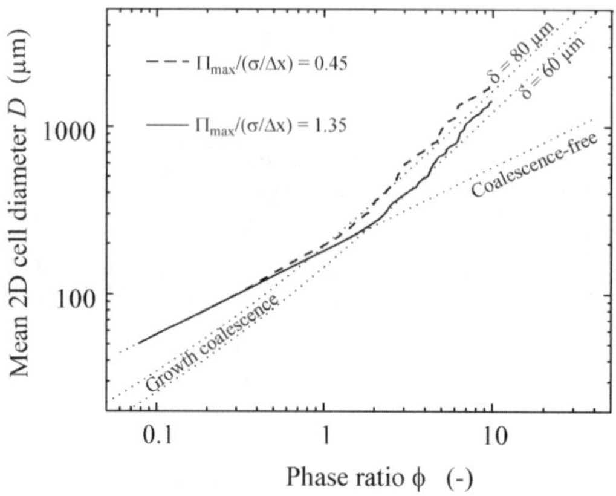

Figure 5.12: Growth coalescence. Mean cell diameter as a function of the phase ratio for Fig. 5.11. *Dotted lines*: Eq. (5.18) with $\delta = 60$ µm and $\delta = 75$ µm. (2D-LBM simulation, parameters: Sim7 + Sim8, Table C.4)

$$D = \frac{\delta}{\sqrt[m]{\frac{\phi+1}{\phi}} - 1}, \tag{5.18}$$

where $m = 2$ and $m = 3$ in 2D and 3D, respectively.
Equation (5.18) reveals a linear relationship between D and ϕ for $\phi \gg 1$:[13]

$$D \approx m\,\delta\,\phi. \tag{5.19}$$

From the last equation the expansion coefficient for this growth mode is determined to be $n = 1$. Thus, a constant mean material thickness leads to a linear relationship between the mean cell diameter and the phase ratio. For high disjoining pressures, Plateau borders are suppressed and δ can be identified with the range of the disjoining pressure and the mean cell wall thickness. If the disjoining pressure is low, pronounced Plateau borders develop and the mean material thickness rises, see Fig. 5.12.

Growth coalescence is sometimes obscured by avalanche-like coarsening processes of the cell structure, which are manifested by a step-like increase of the mean cell diameter, see Figs. 5.13 and 5.12.[14] Every step represents an avalanche event with

[13]Note that this is exactly the relationship which governs the evolution of metal foam produced by quite different processes [157]. This correlation was first described in [147].

[14]In dry aqueous foams, cell ruptures are observed to be correlated and to form avalanches [184].

widespread regions of the foam strongly coarsening in a cascade-like way. After this event, the mean cell diameter approaches the functional dependence given by Eq. (5.18) until the next avalanche event occurs.

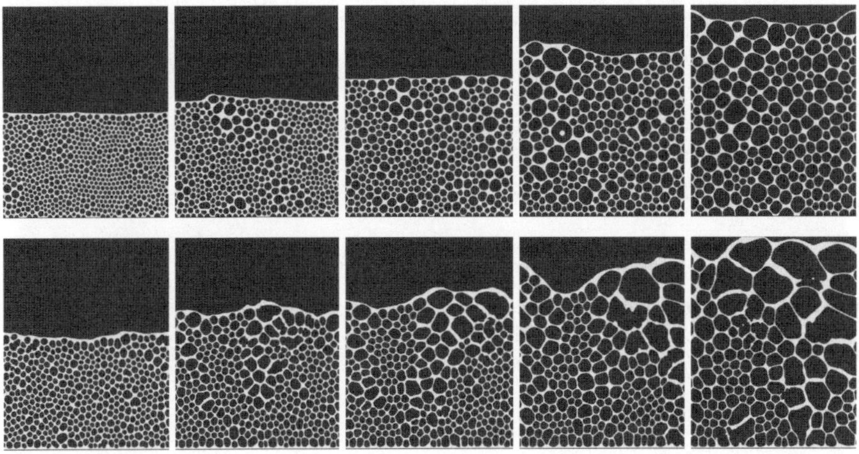

Figure 5.13: Avalanches. Foam coarsening is not homogeneous but takes place locally in a cascade-like way. *Top row*: Small Ca and $\frac{\Pi_{\mathrm{max}}}{\sigma/\Delta x} = 0.45$. *Bottom row*: Large Ca and $\frac{\Pi_{\mathrm{max}}}{\sigma/\Delta x} = 0.9$. (2D-LBM simulation, parameters: Sim9-Sim10, Table C.5)

5.3.3 Influence of the Nuclei Density

During all expansion stages, non-stabilized evolution is governed by the initial nuclei density. Do we expect the same behavior in case of stabilization? In the following, we will see that the presence of stabilization and the corresponding length scale change this behavior.

Figure 5.14 shows the influence of the initial nuclei density on the evolution of the mean cell diameter for stabilized evolution.

For low relative densities, the mean cell diameter follows coalescence-free expansion which is proportional to $1/\sqrt[m]{n_0}$. The higher the nuclei density is the earlier changes the behavior from coalescence-free growth to growth coalescence. The mechanism change is quite abrupt and very close to the intersection point of the theoretical curves. To increase the nuclei density is only efficient as long as foam evolution is coalescence-free. For higher phase ratios, the mean cell diameter is independent of the initial nuclei density. In this regime, the stabilization mechanism determines the

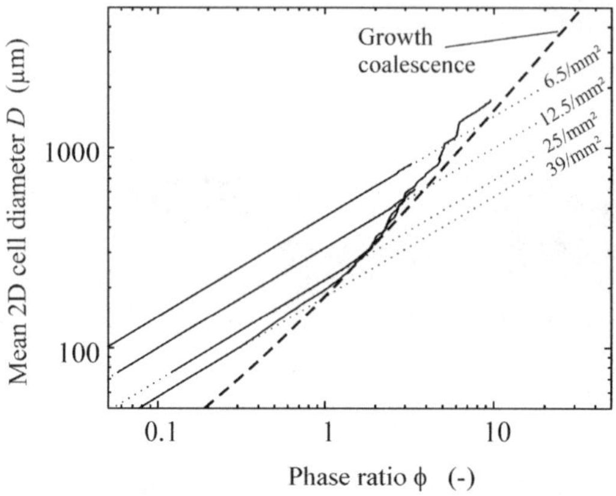

Figure 5.14: Influence of the nuclei density on the evolution of the mean cell diameter for stabilized expansion. *Full lines*: Simulation. *Dotted lines*: Coalescence-free expansion with $\delta = 80\ \mu$m. *Dashed line*: Growth coalescence. (2D-LBM simulation, parameters: Sim7, Table C.4)

relationship between cell diameter and relative density. Generally, the transition from coalescence-free evolution to growth coalescence takes place at higher phase ratios than that from coalescence-free to non-stabilized evolution.

5.3.3.1 Critical Nuclei Density

How high has the nuclei density to be in order to achieve the lowest possible mean cell diameter for a particular relative density? In the following, this nuclei density is designated as critical nuclei density n_0^{crit}. The critical nuclei density is not a constant but a function of the final phase ratio and the mean material thickness, $n_0^{\mathrm{crit}} = n_0^{\mathrm{crit}}(\phi, \delta)$, and decreases with increasing phase ratio. We calculate n_0^{crit} from the intersection point of the theoretical curve for coalescence-free expansion, Eq. (5.5), and that for growth coalescence, Eq. (5.19),

$$\sqrt[m]{\frac{2\ m}{n_0^{\mathrm{crit}}\ \pi}\phi} \quad \stackrel{!}{=} \quad m\ \delta\ \phi, \tag{5.20}$$

and resolve for n_0^{crit}:

$$n_0^{\mathrm{crit}} = \frac{2}{\pi\ m^{m-1}\ \delta^m\ \phi^{m-1}} \quad \stackrel{m=3}{\propto} \quad \frac{1}{\delta^3\ \phi^2}. \tag{5.21}$$

Figure 5.15 shows the critical nuclei density in three dimensions for different mean material thicknesses. The critical nuclei density is very sensitive on δ and increases by a factor of 8 if δ is halved. The shaded area indicates nuclei densities and phase ratios which are characteristic for IFM. The mean material thickness for IFM is somewhere between 50 and 100 μm. Especially for high phase ratios, the nuclei densities observed for IFM (see Sect. 3.2.2) are already sufficient and the nuclei density does not represent some kind of limitation.

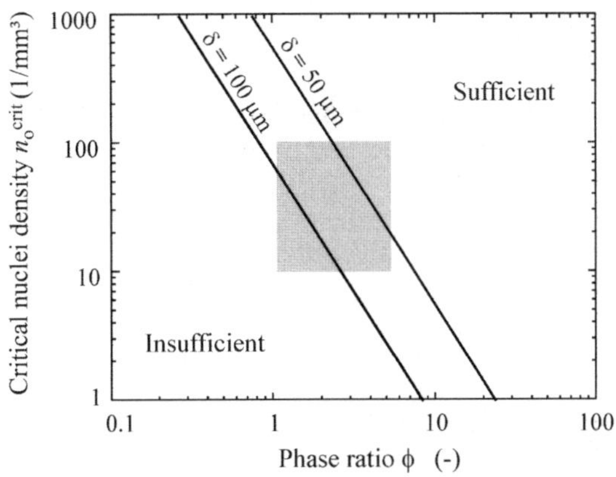

Figure 5.15: Critical nuclei density for stabilized expansion in 3D for δ = 50 μm and δ = 100 μm. The shaded area indicates the region characteristic for IFM.

5.4 Synopsis

In view of the application of the theoretical results on experimental findings the summary is restricted to 3D.
Three different evolution modes have been identified:

1. Coalescence-free evolution in 3D
   This expansion behavior characterizes expansion with unbounded stabilization and an evolution exponent of $n = 1/3$. In addition, it describes the initial expansion behavior if the system is not stabilized or stabilized with upper bound. In all circumstances, the initial nuclei density n_0 can be determined by fitting experimental data with:

$$D = \sqrt[3]{\frac{6}{n_0\,\pi}\,\phi}. \tag{5.22}$$

2. Non-stabilized evolution in 3D
   Non-stabilized evolution is – analogous to coalescence-free evolution – characterized by a direct dependence on the nuclei density which provides the only length scale available. In addition, the mean cell diameter goes to infinity while trying to approach the maximum phase ratio $\phi_{\max}$:

$$D = \frac{c_{-1}}{\sqrt[3]{n_0}} \frac{1}{\sqrt[3]{\frac{\phi+1}{\phi}} - 1 - c_0} \overset{\phi \gg 1}{\approx} \frac{c_{-1}}{\sqrt[3]{n_0}\, c_0} \frac{\phi}{\phi_{\max} - \phi} \tag{5.23}$$

   with $\phi_{\max} = 1/(3\, c_0)$. An evolution coefficient can not be identified for non-stabilized evolution.

3. Stabilized evolution in 3D with upper bound: Growth coalescence
   This evolution mode is characterized by a constant mean material thickness δ:

$$D = \frac{\delta}{\sqrt[3]{\frac{\phi+1}{\phi}} - 1} \overset{\phi \gg 1}{\approx} 3\, \delta\, \phi. \tag{5.24}$$

   The evolution exponent approaches $n = 1$ with increasing ϕ.

An overview is given in Table 5.1.

Table 5.1: 3D Foam evolution behavior for different stabilization modes defined by the disjoining pressure Π.

Disjoining pressure Π	Evolution exponent n	Cell coalescence	Mean material thickness δ
Unbounded	$\frac{1}{3}$	Suppressed	Decreasing
Bounded	1	Due to expansion	Constant
0	Unbounded	Not suppressed	Increasing

It is important to realize that none of the identified evolution laws – stabilized or non-stabilized – is dependent on the viscosity or the surface tension, i. e. the capillary number Ca. This is a result which has not been expected and which contradicts the general conviction that both material properties are of essential importance during foaming.[15]

[15] Please note that the observation that these material properties do not affect the evolution laws does not mean that there is no influence at all. The appearance of the cells is strongly dependent on these material parameters. Nevertheless, the influence on the mean cell diameter is of minor importance.

6 Endogenous Stabilization

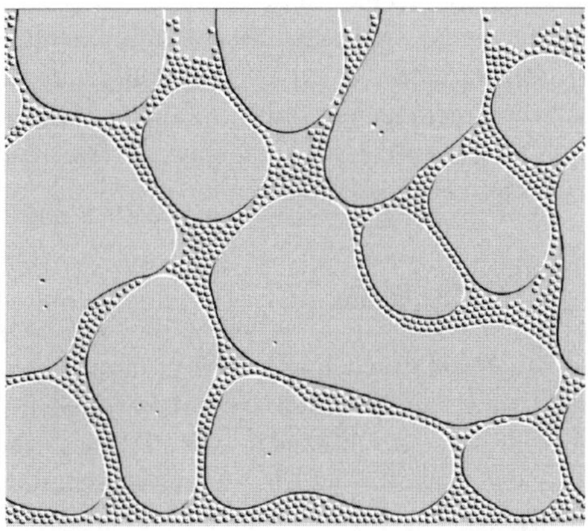

Is there any foam stabilization mechanism present during IFM? At first glance we would say no. Non-conditioned metal melts do not foam! Generally, particles have to be added to metal melts to make them foamable.[1] Since we do not add any particles, any kind of stabilization is unexpected.

Nevertheless, there are particles present during IFM because foam evolution and solidification take place simultaneously. A huge number of primary α-phase particles develops in the semi-solid melt during solidification, see Sect. 2.5. These particles have two effects. On the one hand, the viscosity is strongly increased. This effect leads to some kind of dynamic stabilization, i. e. drainage effects are retarded. On the other hand, the primary particles get captured between neighboring bubbles and represent obstacles which delay or completely stop cell wall thinning. As obstacles, these particles have a static stabilization effect due to particle confinement which we refer to as endogenous stabilization.

[1]See Sect. 4.3.4.

The aim of this chapter is to investigate the role of solid particles during foam evolution with the help of numerical simulation.[2] It is shown that particles confined in cell walls generate a disjoining pressure (Sect. 6.1). Besides the numerical approach and the visualization of the underlying barrier effect against coalescence, the magnitude of the disjoining pressure and its dependence on particle size and particle volume content is estimated with the help of a simple analytical model. Numerical simulation shows that the self-organizing effect of the restricted area in a cell wall leads to layered structures with an oscillatory disjoining pressure. This is the origin of a stepwise cell wall thinning known as stratification (Sect. 6.2).

The disjoining pressure generated by solid phase particles makes it possible to foam metal melts (Sect. 6.3). The resultant foam structures show typical features such as thick walls, local swellings, thermodynamically unfavorable states, etc., which can be traced back to the underlying stabilization mechanism. Thus, stabilization by particle confinement not only explains stabilization during IFM but also reveals the origin of many peculiar structural properties observed in metal foams produced by other foaming methods [146].

6.1 Disjoining Pressure

Particles get captured in cell walls if their diameter surmounts the cell wall thickness. If the particles are wet[3] by the liquid they are not expelled from the wall and the liquid is forced to surround them. Between two particles, the liquid interface is forced to assume a concave curvature which generates a capillary counter pressure, the disjoining pressure Π, see Fig. 6.1.

It is interesting to note that the ideal particle distribution seems not to be one with constant distance between particles. In contrast, the surface tension generates an attractive force between near particles and a repulsive one between far ones. The neck formation is the starting point of cell wall rupture.[4]

In Sect. 6.1.1 we use a 2D-LBM model to calculate the disjoining pressure for different particle volume contents. This section is followed by a simple analytical 3D model, see Sect. 6.1.2, which provides us with an analytical expression for the disjoining pressure and its dependencies in three dimensions.

[2]The details of the numerical approach are described in Part III.

[3]For cubic metals, it appears justified to consider that a pure solid metal is perfectly wet by its own melt [52]. We assume that this is also true for the casting alloys used for IFM.

[4]Below we will see that local formation of a neck is intimately correlated with the fact that the disjoining pressure develops a maximum with decreasing mean cell wall thickness.

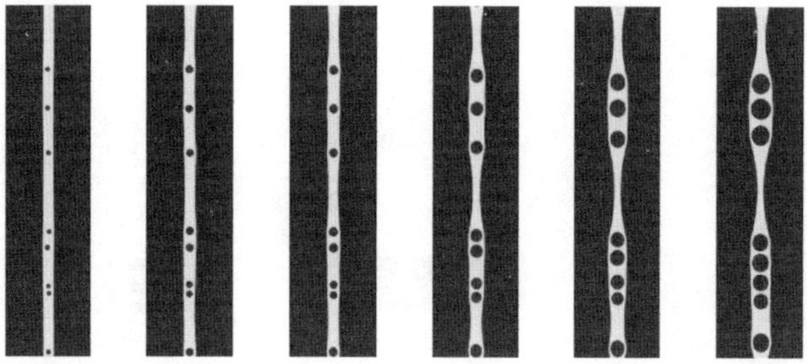

Figure 6.1: Particles confined in a cell wall. From *left* to *right*: Particle diameter D_p is increased. The particles lead to a concave surface which is the origin of the disjoining pressure. There is an attractive interaction between near particles and a repulsive interaction between far ones. (2D-LBM simulation)

6.1.1 2D-LBM Simulation

In order to estimate the disjoining pressure Π generated by particles confined in cell walls it is assumed that they are uniformly distributed along an infinitely long cell wall. In this case, the problem is reduced to a unit cell with only one particle and periodic boundary conditions. In order to estimate Π, a tiny particle is placed into a cell wall and expanded until a diameter of 40 lattice sites is reached, see Fig. 6.2.

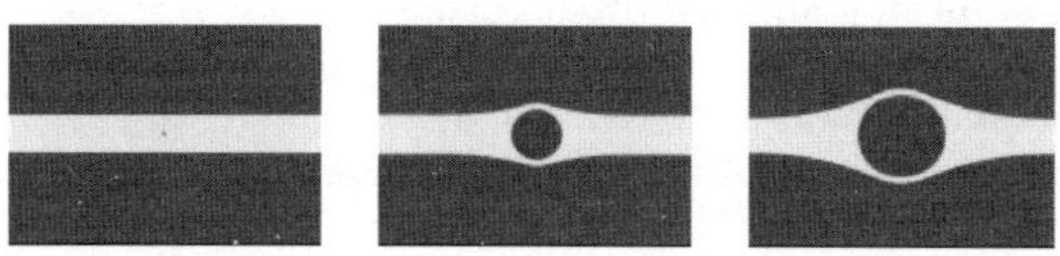

Figure 6.2: LBM simulation of the disjoining pressure. A particle is expanded in a cell wall with constant fluid content until a particle diameter D_p of 40 lattice units is reached. The disjoining pressure is identified with the the pressure in the liquid after equilibration. (2D-LBM simulation, periodic boundary conditions in x-direction)

In order to simulate particles we use degenerated bubbles.[5] These bubbles show a very high surface tension and a high disjoining pressure in order to prevent coalescence. Their effective diameter is about 3 to 6 lattice sites larger than their diameter. After equilibration the mean pressure in the liquid, which is identified

[5]The dynamic behavior of degenerated bubbles is surely different from that of compact particles. Nevertheless, we only want to demonstrate their barrier effect. In this case it does not matter whether particles or degenerated bubbles are used.

with Π, is determined. The results for different relative particle volume contents V_{rel} are depicted in Fig. 6.3.

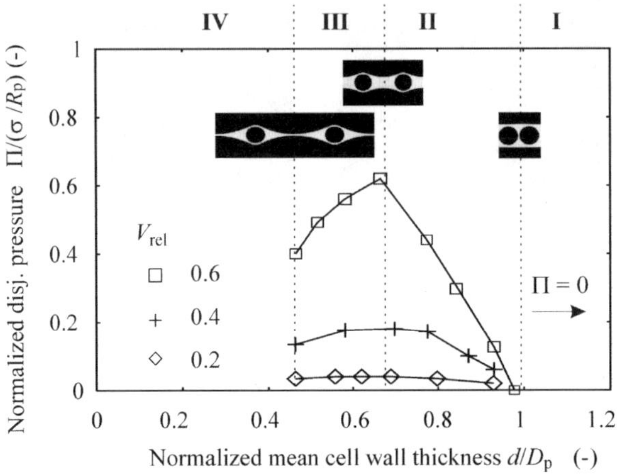

Figure 6.3: Normalized disjoining pressure as a function of the normalized mean cell wall thickness for different relative particle volume contents V_{rel}. (2D-LBM simulation)

As expected, the disjoining pressure increases with increasing particle volume content. Four different domains are distinguished:

- **Domain I:** No effect, $\Pi = 0$.
  The cell wall thickness is larger than the particle diameter and a disjoining pressure does not develop.

- **Domain II:** Stabilization, $\frac{\partial \Pi}{\partial d} \leq 0$.
  The disjoining pressure increases with decreasing mean cell wall thickness d and the cell wall is stabilized.

- **Domain III:** Destabilization, $\frac{\partial \Pi}{\partial d} > 0$.
  The disjoining pressure decreases with decreasing mean cell wall thickness. In this region, cell wall thinning is accelerated.

- **Domain IV:** Cell wall rupture.
  Cell wall rupture takes place when the two opposed interfaces touch each other.

The most surprising result is that the disjoining pressure develops a global maximum far before cell wall rupture occurs. At the moment, when the mean cell wall thickness passes the maximum, destabilization will lead to rapid local thinning followed by cell wall rupture. This is a very interesting result because the mean as well as the local cell wall thickness are not small at the transition between domain II and III. Thus, the transition between stabilization and destabilization until rupture takes place at comparatively thick cell walls.

6.1.2 3D-Analytical Model

The numerical calculation in Sect. 6.1.1 is based on a 2D-LBM simulation. The point is, whether the qualitative progression of the disjoining pressure and, in particular, the maximum also develop in three dimensions. In addition, numerical simulation does us not provide with an analytical expression for the dependence on the parameters such as the surface tension, the particle diameter or the particle volume fraction V_{rel}.

In the following, an analytic expression for Π is derived on basis of a simple 3D model [146]. The particles form obstacles within the cell walls. This obstacle effect is modeled by assuming that the gas–liquid interface is locally pinned if the cell wall thickness reaches the mean particle diameter, see Fig. 6.4.

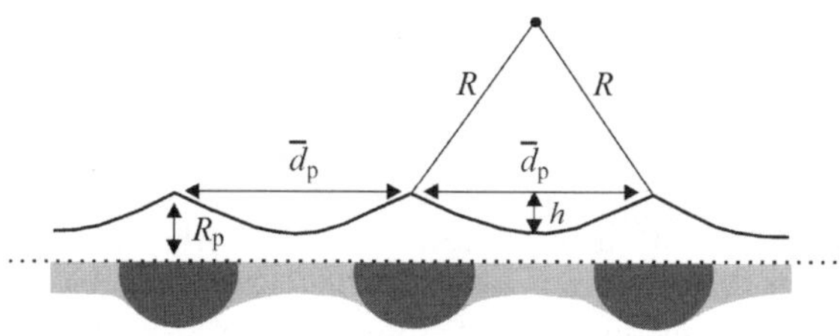

Figure 6.4: Analytical model for the development of the disjoining pressure [146].

Cell wall thinning leads to the development of a concave surface described by the radius of curvature R. As a result, local repulsive forces emerge which act against further cell wall thinning between the particles. In a mean field picture, this local repulsive force can be replaced by a repulsive pressure, the disjoining pressure:

$$\Pi = \frac{2\,\sigma}{R}, \tag{6.1}$$

with the surface tension σ.

The radius of curvature R follows from simple geometric relations,

$$R = \frac{1}{2}\left(\frac{\overline{d}_{\mathrm{p}}^{\,2}}{4h} + h\right), \tag{6.2}$$

where $\overline{d}_\mathrm{p}$ denotes the mean particle distance and h is the local indentation depth of the cell wall.

In order to estimate Π, $\overline{d}_\mathrm{p}$ has to be expressed as a function of the mean particle radius R_p and the relative particle volume fraction V_rel. Assuming that the total wall volume V_tot belonging to one particle is composed out of a disc plus two cones, V_tot may be expressed as

$$V_\mathrm{tot} = \pi \frac{\overline{d}_\mathrm{p}^{\,2}(R_\mathrm{p} - h)}{2} + \frac{\pi \overline{d}_\mathrm{p}^{\,2} h}{6} = \frac{\pi \overline{d}_\mathrm{p}^{\,2}(R_\mathrm{p} - \frac{2}{3}h)}{2}. \qquad (6.3)$$

The relative volume follows from the particle volume $V_\mathrm{p} = \frac{4}{3}\pi R_\mathrm{p}^3$ and V_tot:

$$V_\mathrm{rel} = \frac{V_\mathrm{p}}{V_\mathrm{tot}} = \frac{8}{3} \frac{R_\mathrm{p}^3}{\overline{d}_\mathrm{p}^{\,2}(R_\mathrm{p} - \frac{2}{3}h)} \qquad \text{and} \qquad \overline{d}_\mathrm{p}^{\,2} = \frac{8}{3} \frac{R_\mathrm{p}^3}{V_\mathrm{rel}(R_\mathrm{p} - \frac{2}{3}h)}. \qquad (6.4)$$

Combining Eqs. (6.1), (6.2) and (6.4) gives the disjoining pressure:

$$\Pi = \frac{2\sigma}{R} = \frac{16\, h\, \sigma}{\frac{8R_\mathrm{p}^3}{3V_\mathrm{rel}(R_\mathrm{p} - \frac{2}{3}h)} + 4h^2} = \frac{\sigma}{R_\mathrm{p}} \frac{6\,(1 - \overline{d})}{\frac{2}{3V_\mathrm{rel}\overline{d}} + \frac{9}{4}(1 - \overline{d})^2}, \qquad (6.5)$$

where the normalized mean cell wall thickness $\overline{d}$ is defined by $\overline{d} = \frac{d}{2R_\mathrm{p}} = \frac{d}{D_\mathrm{p}}$ and $V_\mathrm{tot} = d\pi \frac{\overline{d}_\mathrm{p}^{\,2}}{4} = \frac{\pi \overline{d}_\mathrm{p}^{\,2}(R_\mathrm{p} - \frac{2}{3}h)}{2}$. The mean cell wall thickness is given by $d = 2(R_\mathrm{p} - \frac{2}{3}h)$.

Equation (6.5) reveals the dependencies of the disjoining pressure. The most obvious dependence is that on the particle radius R_p which determines the scale. The actual value of the pressure is given by the particle volume fraction and the normalized mean cell wall thickness. Figure 6.5 shows the normalized disjoining pressure as a function of the normalized mean cell wall thickness $\overline{d}$ for different relative particle volume contents.

The results of the 3D model are completely consistent with the 2D-LBM simulation, see Fig. 6.3. Qualitatively, the same functional dependence is predicted. There are some quantitative differences with respect to the dependence on the particle volume content and the position of the maximum. In two dimensions, the dependence on the particle volume content is much more pronounced than in three dimensions where already low particle volume contents lead to quite high disjoining pressures.[6] In two dimensions, the influence of the particle volume content on the height of the maximum is about quadratic while there is a nearly linear relationship in three dimensions. In addition, the maximum of the disjoining pressure is shifted to somewhat lower values.

[6] This is an important result. Stabilization is already quite high for low volume fractions of solid phase where the increase of the viscosity is yet quite low.

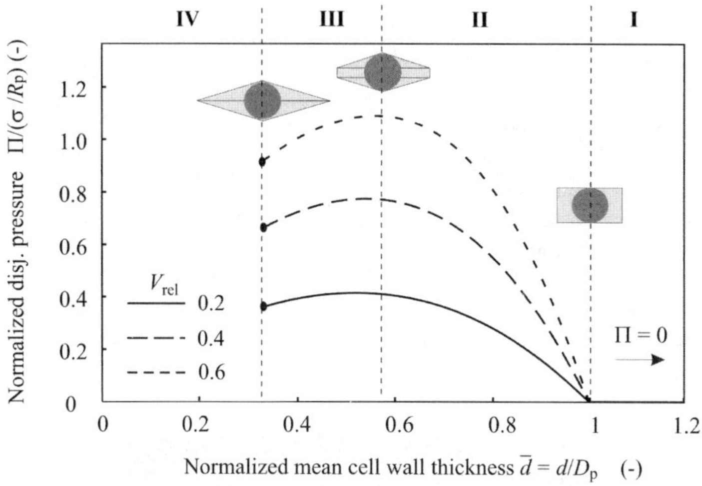

Figure 6.5: Disjoining pressure as a function of the mean cell wall thickness for different particle volume contents [146]. (3D-analytical model)

If we compare the magnitude of the disjoining pressure with the Laplace pressure at the Plateau borders of a foam it becomes clear that Π is rather high. In a stable foam, the sucking pressure of the Plateau borders balances the disjoining pressure, see Eq. (4.27). The curvature of the Plateau borders is thus of the same order of magnitude as $1/R_p$. Consequently, the Plateau borders are predicted to assume a smaller and smaller radius of curvature with increasing particle volume content. In addition, particle stabilization implies the existence of stable foam structures with very small Plateau borders where nearly all material is located in the cell walls.[7]

An additional implication of Fig. 6.5 is, that the mean cell wall thickness d is determined by the particle diameter:

$$0.5\, D_p \leq d \leq D_p. \tag{6.6}$$

Thinner cell walls are highly unstable, i. e. cell walls are yet quite thick at rupture. Again, this observation is a quite familiar observation for metal foams. In addition, the lower limit of the mean cell wall thickness is a measure of the radius of the particles responsible for stabilization.[8]

[7] Compare with Sect. 4.3.4.

[8] There are also stable cell walls with $d > D_p$, see Sect. 6.2.

The transition from stable to unstable states is shifted to thinner cell walls if the particle volume fraction is decreased. On the other hand, a lower particle volume fraction is associated with a lower disjoining pressure, i. e. a lower stabilization.

6.2 Stratification

Up to now, the basis for the calculation of the disjoining pressure has been a unit cell with only one particle. As a result, the disjoining pressure is zero if the mean cell wall thickness exceeds the particle diameter. However, in the presence of many particles and a high particle volume fraction the action of the confined area of a cell wall becomes noticeable much earlier. In the following, the self-organizing effect in the restricted area of a cell wall and the resulting disjoining pressure is investigated, see Fig. 6.6.[9]

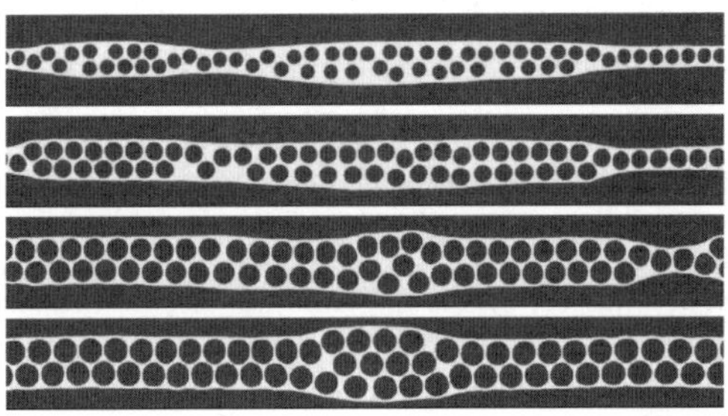

Figure 6.6: Self-organization of particles in the restricted volume of a cell wall. From *bottom* to *top*: Particle diameter and particle volume fraction increase. (2D-LBM simulation)

The restricted area of the cell wall in combination with the restoring force due to the surface tension forces enclosed particles to assume ordered structures. From Fig. 6.6 the tendency for a layered particle arrangement with one, two or three layers becomes visible. Ordered structures minimize the surface curvature and thus the disjoining pressure. The resulting cell walls show a non-uniform thickness with abrupt thickness transitions which are stepwise with a height of about the particle diameter.

For the numerical simulation of the disjoining pressure unit cells with one, two and three particles are utilized. Their equilibrium arrangement and the resulting disjoining pressure are depicted in Fig. 6.7.

[9]See also Sect. 4.3.3.3.

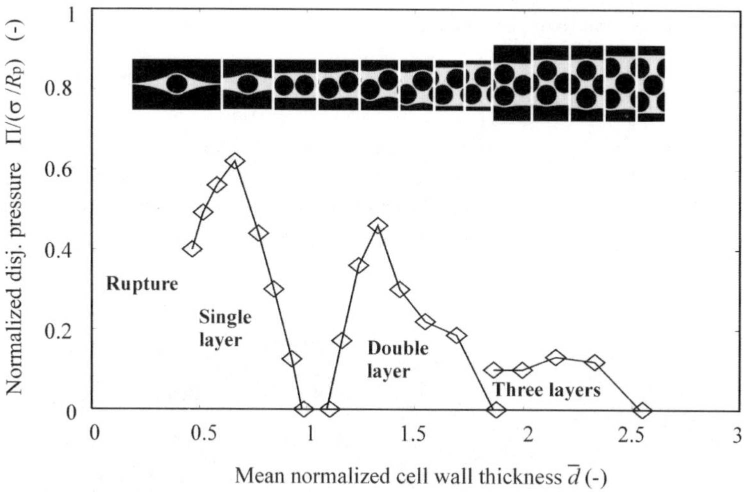

Figure 6.7: Stratification. Normalized disjoining pressure and particle arrangement as a function of the normalized mean cell wall thickness. (2D-LB simulation, particle volume fraction $V_{\text{rel}} = 0.6$)

The most important insights are as follows:

- Π oscillates and its range is much larger than the particle diameter.
- The distance of the maxima is about the particle diameter.
- The height of the maxima strongly decreases with increasing mean cell wall thickness.

These observations clearly show that the stability of cell walls is influenced by their mean thickness. The system will organize itself in such a way that stable configurations are assumed. Thus, we expect to observe preferably cell walls with thicknesses of about one, two or three particle diameters. In addition, abrupt thickness transitions along stable cell walls are predicted.

6.3 Foam Evolution

In the preceding sections, the barrier effect of particles and the nature of the resulting disjoining pressure in isolated cell walls has been treated. The aim of this section is to demonstrate the stabilizing effect of particles during foam evolution. Besides the stabilizing effect, the characteristic structural features involved with this kind of stabilization mechanism are discussed.

6.3.1 Evolution Laws

The barrier effect of particles makes non-foaming melts foamable. Figure 6.8 shows simulated 2D cell structures for two different particle volume contents but identical initial conditions. The stabilization effect increases with increasing particle content. The majority of the cell walls consists of one, two or tree particle layers, compare with Sect. 6.2.

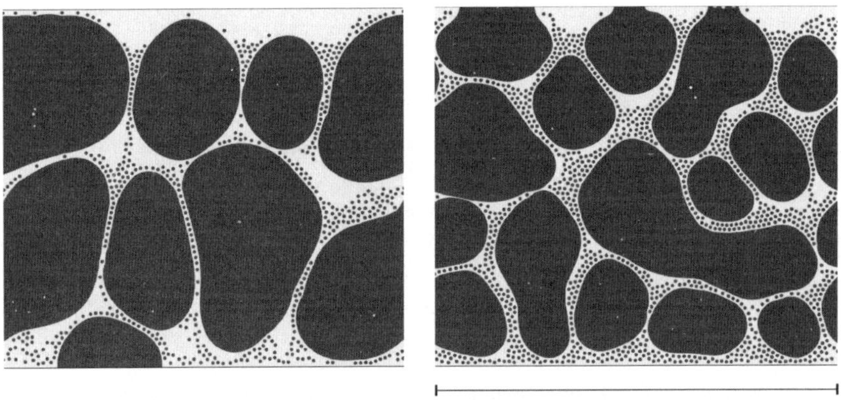

Figure 6.8: Cellular structures stabilized by particles after 300 ms. *Left*: $V_{\mathrm{rel}} = 34\%$ (70 nuclei, 475 particles). Right: $V_{\mathrm{rel}} = 68\%$ (70 nuclei, 950 particles). (2D-LBM simulation, effective particle radius ≈ 25 µm, parameters: Endo1-Endo2, Table C.6)

The stabilization effect of the particles is rather low as long as the phase fraction ϕ is small. During this initial phase, particle confinement is not possible since the particles are still able to escape from an evolving cell wall, see Fig. 6.9.
The situation changes when the phase fraction increases and $\phi > 1$. In this case, the particles get trapped in cell walls where they develop their barrier effect. The stabilizing effect of the particles emerges only for higher phase ratios whereas bubble coalescence is virtually not hampered at low phase ratios.

Figure 6.10 shows the evolution of the normalized mean cell diameter for the two examples in Fig. 6.8.

For small ϕ, expansion is coalescence-free and the evolution exponent $n = 1/2$. Variations from coalescence-free expansion occur in the range $\phi = 0.3$–0.4 for both particle volume contents.

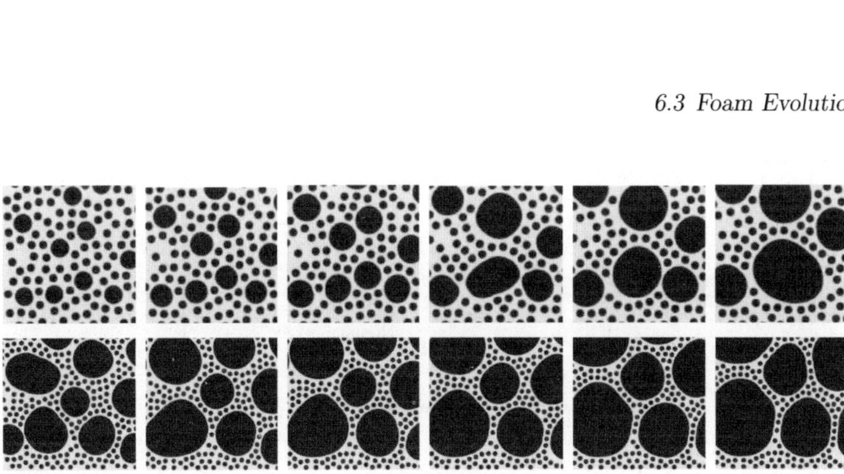

Figure 6.9: Influence of the phase fraction ϕ on endogenous particle stabilization. *Upper row:* $\phi < 1$. The particles get expelled from transient cell walls during bubble growth. *Lower row:* $\phi > 1$. The particles get confined in transient cells walls and develop the barrier effect. (2D-LB simulation)

During further expansion, the mean cell diameter follows non-stabilized evolution, Eq. (5.14), as long as the phase ratio is lower than a critical value, ϕ_{crit}, which designates the point where the stabilizing effect of the endogenous particles becomes apparent. The critical phase ratio ϕ_{crit} is not a constant but decreases with increasing particle volume fraction. Stabilization is not very pronounced for low particle volume contents since the mobility of the particles is still very high. Things are different for higher particle volume contents. In this case, the mobility of a particle is strongly restricted due to the presence of the other ones and the barrier effect develops much earlier. As a result, the transition between non-stabilized expansion and stabilized expansion is at $\phi_{\text{crit}} = 1.5$ and $\phi_{\text{crit}} = 0.4$ for $V_{\text{rel}} = 34\,\%$ and $V_{\text{rel}} = 68\,\%$, respectively.

Eventually, the curve progression approaches growth coalescence, Eq. (5.18), with an evolution coefficient $n = 1$ for $\phi > \phi_{\text{crit}}$. For the high particle volume content, the mean material thickness δ for stabilized expansion is determined to be about 50 μm, which is twice the effective particle diameter. This value increases to 70 μm if the effective particle volume content is lower.

In summary, the effect of endogenous particles on the foaming behavior of a liquid is characterized as follows. At low phase ratios, there is no stabilizing effect apart from an effective increase of the viscosity of the liquid. Expansion starts without coalescence events and changes to non-stabilized expansion at about $\phi = 0.3$–0.4. At higher phase ratios, i. e. $\phi > \phi_{\text{crit}}$, the particles start to act as mechanical barriers and lead to stabilized expansion characterized by growth coalescence. Figure 6.11 shows a schematic of the expansion behavior.

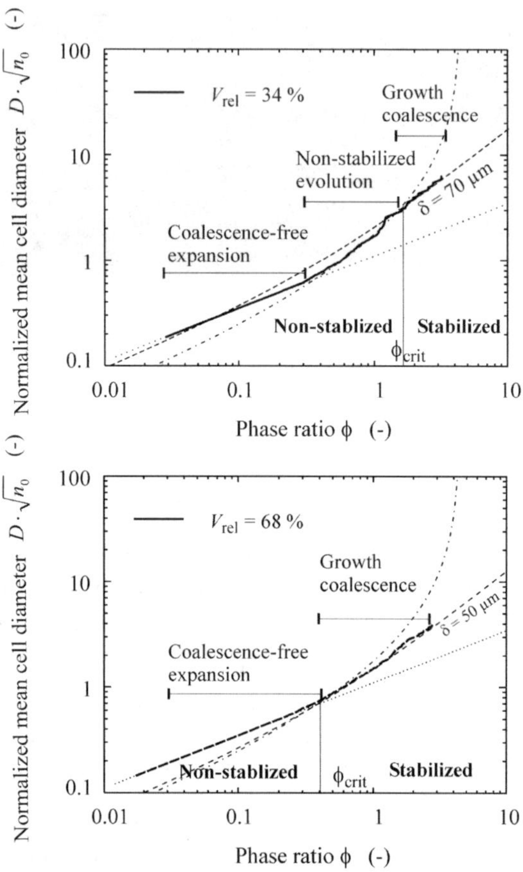

Figure 6.10: Normalized mean cell diameter as a function of the phase ratio for the two examples in Fig. 6.8. *Top*: $V_{\rm rel} = 34\,\%$ (475 particles). *Bottom*: $V_{\rm rel} = 68\%$ (950 particles). Non-stabilized evolution (*dashed-dotted line*): Eq. (5.14) with $c_{-1} = 0.55, c_0 = 0.105$. Stabilized evolution (*dashed line*): Eq. (5.18). (2D-LBM simulation, parameters: Endo1-Endo2, Table C.6)

The schematic suggests an additional region 3 where expansion is coalescence free due to stabilization. The existence of this expansion behavior, which is expected to occur for low initial nuclei densities, could not be satisfactorily confirmed by numerics due to strong variations caused by the too small system size. Nevertheless, this behavior is observed in experiments, see Sect. 3.2.

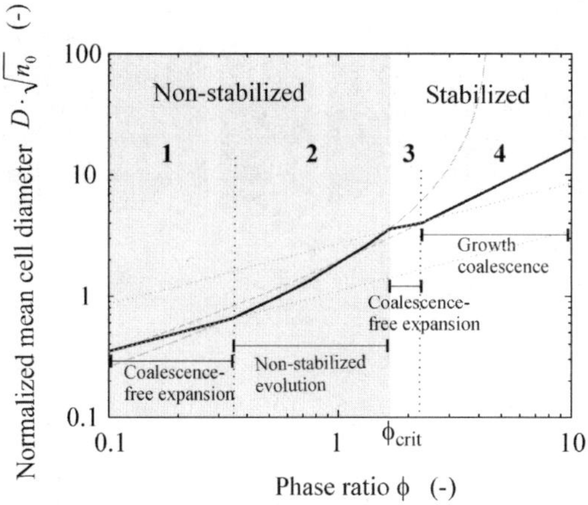

Figure 6.11: Evolution behavior for foams stabilized by endogenous particles. Non-stabilized evolution. (**1**) Coalescence-free expansion, (**2**) non-stabilized evolution, stabilized evolution, (**3**) Coalescence-free expansion, (**4**) Growth coalescence.

6.3.2 Cell Wall Thinning and Rupture

An important process during foam evolution is cell wall thinning which is mainly due to a stretching effect during foam expansion. If the volume of the foam increases cell walls have to get thinner. Figure 6.12 shows the thinning of a cell wall in the presence of particles. As long as there are multiple-particle layers present, the particles redistribute to form a mono-particle layer.

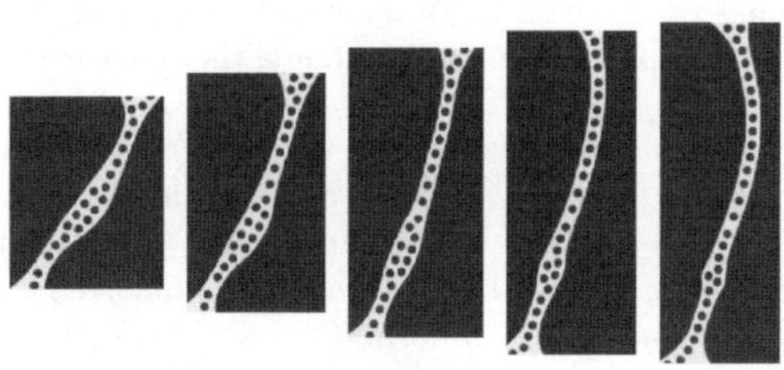

Figure 6.12: Cell wall thinning in the presence of particles. (2D-LBM simulation)

It is worth to mention that the thickness of the cell wall is determined by the particle diameter. There are local swellings indicating the presence of multiple-particle layers. The cell wall does not develop the typical form with a minimum thickness in the middle.

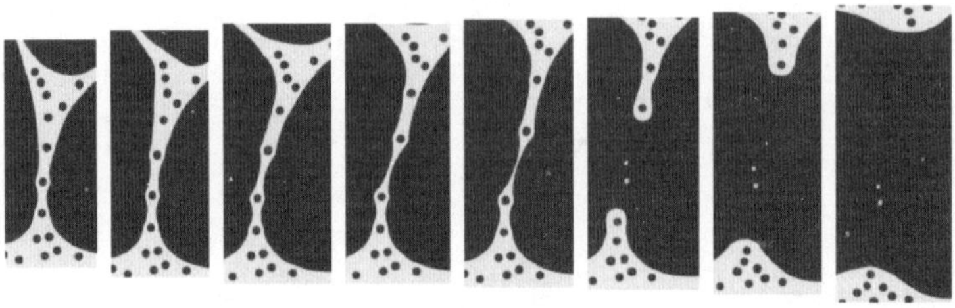

Figure 6.13: Cell wall rupture in the presence of particles. (2D-LBM simulation)

Cell wall rupture is initiated when the cell wall thickness falls locally below the particle diameter. Figure 6.13 shows a rupture event which is induced by foam expansion. Rupture is characterized by a local thinning process while the rest of the cell wall keeps approximately its thickness. At rupture, the mean thickness of the cell wall is about the particle diameter. This behavior is different from the thinning of transient cell walls with a continuous thickness decrease and the rupture point in the middle of the wall.

6.4 Synopsis

The characteristic features of foams stabilized by (endogenous) particles can be summarized as follows:

- Generation of a disjoining pressure with upper bound
  Particles generate a disjoining pressure due to their obstacle effect at high phase ratios since they get captured within cell walls. The amplitude and range of the disjoining pressure strongly increases with increasing particle volume fraction. For higher particle volume contents, the range is about three particle layers.

- Existence of a lower limit for the mean cell wall thickness
  Cell walls are highly unstable if their mean cell wall thickness falls below a critical value which is determined by the particle diameter. Thinner cell walls are hardly present and mainly in form of local indents. The particle diameter determines the material thickness. In order to produce high porous foams with small cell diameter, the particle diameter has to be as small as possible.

- Stratification: Step-by-step cell wall thinning
  Cell walls do not thin in a continuous way but by stepwise transitions. The height of the steps corresponds to the particle diameter and is nearly

constant. The reason for stratification is the oscillating behavior of the disjoining pressure.

- Extended stable cell walls with thickness fluctuations
  The presence of the particles leads to stable cell walls. These cell walls either have a more or less constant thickness or show stable local indents or swellings.

- Existence of peculiar cell forms and arrangements
  For high particle volume contents, the disjoining pressure is in the same order of magnitude as the capillary pressure at the Plateau borders. The sucking effect of the Plateau is completely balanced. The barrier effect of the particles converts highly unstable cell arrangements with a very high surface energy to stable ones.

- Evolution exponent
  The evolution exponent n approaches $n = 1$ with increasing phase ratio and the system shows the phenomenon of growth coalescence. The foaming behavior is analogous to that of a liquid with a disjoining pressure which is bounded above.

Is it reasonable to describe the effect of endogenous particles with the help of a disjoining pressure? The stabilizing effect of endogenous particles is based on a barrier effect which depends on the local situation of the cellular structure. Stabilization develops if the mobility of the particles is restricted due to cell wall confinement for high phase ratios. Obviously, a disjoining pressure is not able to describe the entire expansion behavior of foams stabilized by endogenous particles. Nevertheless, if stabilization is high and the phase ratio not too low, the concept of a disjoining pressure in fact describes the underlying expansion behavior very well.[10]

[10]See also Sect. 7.2.

Part III
NUMERICAL SIMULATION

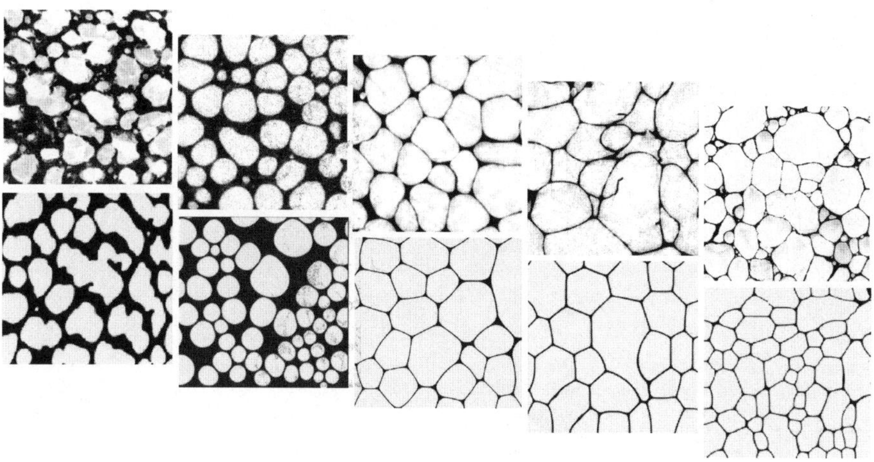

Numerical Simulation

The purpose of computing is insight, not numbers.

R.W. Hamming

The purpose of computing is insight, not pictures.

L.N. Trefethen

Foam evolution is characterized by a huge internal fluid–gas interface which strongly evolves with time. Time evolution is accompanied by cell wall rupture events possibly inducing avalanche-like coalescence events and topological rearrangements of the whole cell structure [57, 58, 233]. These complex processes are the outcome of the interplay of comparatively simple physical phenomena: fluid dynamics, capillary forces and foam stabilization which either can be a surface or a volume effect.

Comprehensive approaches to model aqueous foams with computational fluid dynamical (CFD) methods have not been applied in the past. Time scales (μs) and length scales (nm) seemed to be too short to be numerically resolved. Fortunately, metal foams show many orders of magnitude larger length scales than aqueous foams making them accessible to numerical modeling.

The huge and strongly evolving interface makes the numerical treatment with standard CFD methods very difficult. There are two fundamental problems which arise due to the presence of the sharp fluid–gas interface. Firstly, solving of the Navier-Stokes equations (NSE) is difficult since the derivatives of the macroscopic fields are not continuous at the interface. Secondly, the movement of the interface itself is a formidable task. The lattice Boltzmann method (LBM), which is a mesoscopic approach, has shown to be especially appropriate for the underlying foam formation problem [45, 154, 155, 160, 161, 162, 163].

Now, we are able to simulate comprehensive foam evolution processes in two and three dimensions including nucleation, bubble expansion, aging and eventually decay. Since the employed physical model comprises the governing physics, the outcome of the simulation shows the variety of complex phenomena generally experimentally observed: coarsening, drainage, self-organization in confined geometries, avalanches, etc., without further model assumptions.

Three dimensional simulations are much more complex and time consuming than two dimensional ones. Thus, the system size is strongly restricted for three dimensional simulations. That is one reason why we use exclusively two dimensional simulations for the derivation of the foam evolution laws in Part II. The other reason to use two dimensional simulation is that gas diffusion is up to now only implemented for 2D. Thus, the description of the LBM for foam evolution is restricted to two dimensions although the generalization to three dimensions is possible and has also partly been performed [155, 162].

7 Theoretical Approach

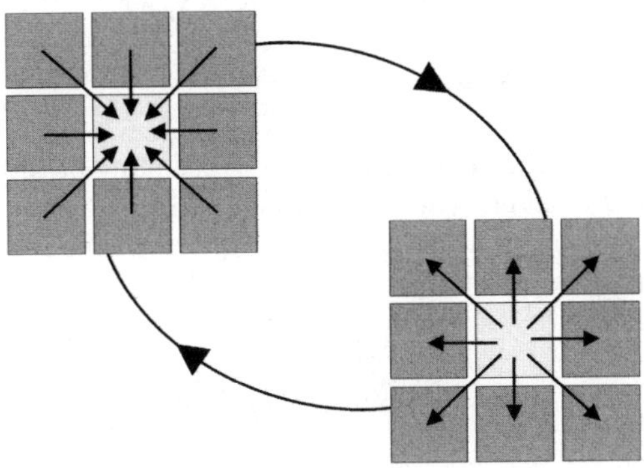

I am conscious of being only an individual struggling weakly against the stream of time.

Ludwig Boltzmann

Numerical simulation of foam structure evolution processes is a formidable task! So far, only some aspects of foam evolution could be described by theoretical approaches. Thus, most of the complex phenomena such as avalanches, rapid structure changes, bubble sorting [120], etc., have not or only hardly been accessible to date.

A short state-of-the-art of foam modeling approaches in literature, see Sect. 7.1, illustrates that there are no methods available which allow the simulation of foam evolution including nucleation, growth, coalescence, coarsening and eventually decay. Generally, numerical or analytical models focus on a particular phenomenon such as foam drainage or coarsening. In addition, the applied models are always associated with strong simplifications of the underlying physical problem to make them solvable. Altogether, the analytical and numerical approaches available in literature are very limited and – so far – only part of the experimentally observed phenomena are captured.

An important step toward a comprehensive simulation model is to identify the governing physical mechanisms responsible for almost all phenomena occurring during foam evolution. For this reason, the selection of the underlying physical model represents the basis for the success of the theoretical approach, see Sect. 7.2.

A further prerequisite for the success of numerical simulation is the availability of an efficient numerical tool in order to solve the underlying physical model represented by a set of coupled differential equations. The lattice Boltzmann approach, whose main principles are briefly presented in Sect. 7.3, shows specific properties which prove especially suitable for complex systems such as foam evolution.

7.1 State-of-the-Art

This section is thought to give a brief review of the state-of-the-art of foam modeling. A comprehensive numerical model, where the whole foam evolution process is investigated, is not known so far. A crude overview of the variety of phenomena involved with foaming is depicted in Fig. 7.1.

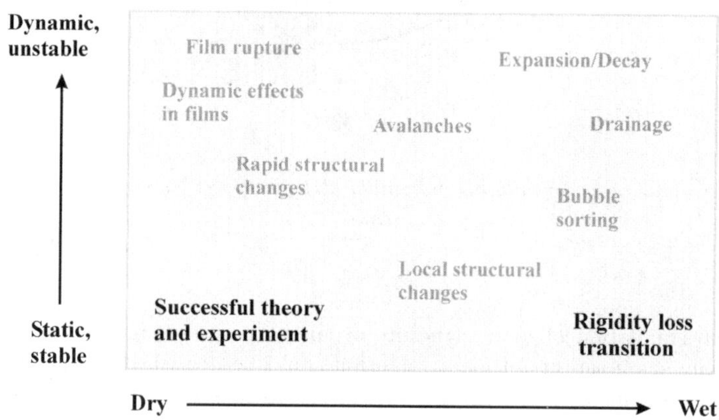

Figure 7.1: Foam phenomena (after Weaire [229, 230])

Whereas a successful theory has been developed in the dry and static regime during the last two decades, there is an obvious lack of understanding in the wet and dynamic region, where metal foams belong to. Up to now, modeling of foam has always concentrated on specific aspects [232], e. g. equilibrium structures, coarsening, drainage [78, 218, 231], rheology [56, 126, 142], etc. Common to all models is that the actual problem is strongly simplified due to the complexity of the task.

In principle, there are two completely different modeling philosophies to render the whole problem more accessible:

1. Structure models
   These approaches model the actual foam structure on the cell level, i. e. cells, cell walls, Plateau borders, etc. Generally, the underlying physics is drastically simplified.

2. Continuum models
   The foam is treated as continuum, i. e. the structure is not resolved. The foam is characterized by a set of state variables, e. g. the mean Plateau border radius, the mean cell size or the mean wall thickness, which obey macroscopic differential equations. These state variables are averages of a volume which is small compared to the system size and large compared to the bubble size [21, 42, 60, 95, 188].

Both approaches are reasonable. Structure models have to be replaced by continuum models for large systems[1] due to the high amount of operating expense. In this section, the focus is on structure models which describe the foam evolution on the cell level.

7.1.1 Evolution of Cellular Structures

Comprehensive reviews of models describing the kinetics of cellular structures are given by Stavans [215] and Glazier and Weaire [89]. In the following, we briefly describe four different approaches which successfully describe aspects of foam evolution behavior. The physical problem is always drastically simplified.

- Vertex models
  The main characteristic of vertex models is that all dynamics is projected on the vertices which obey equations of motion including dissipation and surface tension.These models have been successfully applied on foam rheology investigations including topological changes of the network structure and avalanche processes [37, 76, 192].

- Direct methods
  The philosophy of direct methods is to treat vertices and whole boundaries (or elements thereof) as the basic variables which are relaxed subject to the constraint of $2\pi/3$ at vertex angle (in two dimensions) [28, 136, 231].Topological changes, e. g. T1- or T2-processes, are accessible with this approach, see Fig. 7.2. Numerical studies of structural changes in 2D fluid foams using a hybrid lattice gas model are presented in [221].

[1] Unfortunately, real parts are nearly always large.

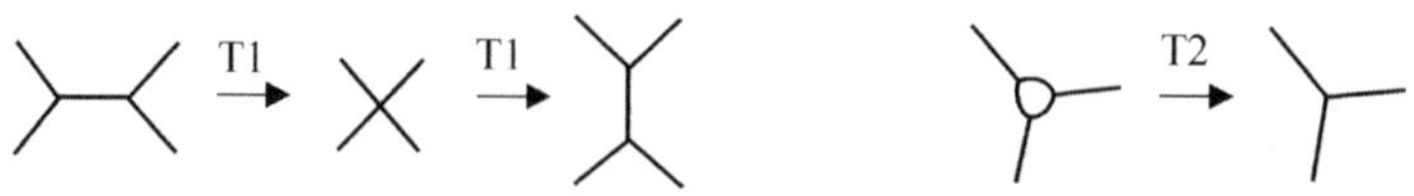

Figure 7.2: Topological transformations in 2D [27, 89]. *Left*: T1-process. *Right*: T2-process.

- Bubble–bubble interaction models
  Bubble-bubble interaction models focus on entire bubbles rather than films or vertices [57, 58, 77]. The models are constructed by pairwise interactions between neighboring gas bubbles characterized by the center positions and radii. Details of the bubble shapes or the motion of the fluid phase are not taken into account. These models comprise melting transitions through decreasing the bubble volume fraction where the static shear modulus vanishes. In addition, avalanche-like topological rearrangements and Bingham plastic behavior[2] emerge from these models.

- Mathematical methods
  Recent approaches [198,208] to handle foam expansion and structure evolution use mathematical models with a high degree of symmetry on the basis of unit cells. These models comprise the influence of viscosity and surface tension but full hydrodynamics and structural changes are not included.

7.1.2 Drainage

Besides structure evolution, drainage, i. e. the fluid flow through or out of a foam structure under the influence of gravitational or capillary forces, is intensely investigated in literature [29, 80, 82, 118, 119, 137, 138]. Generally, the complex problem is reduced to a differential equation, the so called *drainage equation* [226]. The flow is considered in network-like Plateau borders and is normally assumed to be Poiseuille-like. Asymptotic analytical and numerical solutions of the drainage equation are discussed.

7.1.3 Comprehensive Modeling Approaches

All approaches briefly discussed in the preceding paragraphs have one thing in common, they focus on specific aspects of foam – structure or rheology or drainage,

[2] A Bingham plastic is a material that behaves like a rigid body at low stresses but flows like a viscous fluid at high stresses.

etc. Concentration is realized by simplifying the underlying physical model until it turns out to be numerically solvable. Often, the phenomenon, which is aimed at, is included in the model in an artificial way.

Nevertheless, the variety of complex phenomena observed in foams is based on very simple physics, hydrodynamics driven by capillary forces. Phenomena such as drainage, avalanches, local structural changes, bubble sorting, etc., should emerge in a natural way from simulation if full hydrodynamics and capillary forces are taken into account.

The first approach to a comprehensive model was by Körner et al. [45, 154, 155, 158, 159, 160, 161, 162, 163]. On the basis of a cellular automation model, the full hydrodynamics including surface tension and foam stabilization is solved. Phenomena like avalanche-like coalescence, drainage, structural changes, cell deformation, structural evolution in a confined geometry, etc., emerge without further assumptions.

A comprehensive finite element approach to simulate the expansion of gas bubbles in polymer melts is presented in [23, 31].

7.2 Physical Model

Scientists do not deal with truth; they deal with limited and approximate descriptions of reality.

Fritjof Capra

The first step to formulate a physical model of the real foam world is to realize that the plenty of different phenomena involved with foaming is not a manifestation of a variety of physical phenomena but the result of a complex interplay of hydrodynamics, gas supply and diffusion and surface phenomena. Nothing else.

All observed phenomena are expected to emerge from numerical simulation if these aspects are included. Other physical phenomena like melting, solidification, crack formation or oxidation of the surface, are also important with regard to the final foam structure, but are – from a general point of view – of secondary importance. Thus, the model should contain full hydrodynamics, gas supply and gas diffusion, and surface phenomena. The latter comprises capillary and interfacial forces. Melting and solidification are not taken into account.

7.2.1 Hydrodynamic Approach

Neglecting phase transitions and treating the melt as homogeneous fluid reduces the real problem to a two-phase problem, gas bubbles in a fluid. In a further step, we reduce this two-phase system to a one-phase system with free surfaces where the gas phase is not treated explicitly, see also Sect. 4.1.1. Although the dynamics of the gas is not explicitly taken into account it takes a central role during foaming by blowing agents. Bubble nucleation as well as gas release and gas diffusion have to be described.

7.2.2 Gas Bubbles

Gas bubbles are defined by their gas content and volume. The bubble pressure is given by the ideal gas equation, Eq. (4.3).[3] If the bubble pressure minus the capillary pressure exceeds the fluid pressure, the bubble grows, otherwise it shrinks. The amount of gas in a bubble changes due to gas currents from or to the fluid metal where the blowing gas is dissolved.

7.2.3 Bubble Nucleation

The development of bubble nuclei is another matter entirely. Generally, bubble nucleation will preferably be at defects like oxide inclusions or solid particles, etc.

[3] The gas phase of a foam is in fact governed by the ideal gas law, see [102].

We assume heterogeneous nucleation. A huge number of nuclei is already present at the beginning of foaming. These nuclei are stochastically distributed and formed simultaneously at the beginning of foaming. Although very simple, this nucleation model is consistent with experimental observations where the blowing agent particles have been identified as bubble nucleation sites, see Fig. 3.19.

7.2.4 Blowing Agent

Experimentally, the blowing gas results from the decomposition of a metal hydride, in most cases TiH_2 or MgH_2. The hydrides are added to the metal in form of a fine powder and act as point-like hydrogen releasing sources. The total hydrogen concentration field in the melt is the result of the superposition of the multitude of these point-like sources. If the diffusion velocity of the hydrogen in the melt is low or if the blowing agent particles are quite large, the hydrogen concentration field depends on the exact position of the blowing agent particles. Otherwise, the superposition of the individual concentration fields leads to a more or less homogeneous hydrogen concentration field.

Numerically, the decomposition of the blowing agent is modeled by a homogeneous volume source in the fluid domain, i. e. individual blowing agent particles are not taken into account. This method is obviously reasonable if the diffusion length, see Eq. (4.12), of hydrogen on the time scale of foam formation is much larger than the mean blowing agent particle distance. In case of large blowing agent particles, which, in addition, act as nucleation sites,[4] this approach seems at first glance to be very crude. The blowing agent particles located at the fluid–gas interface will directly release the blowing gas into the gas bubbles. The total effect will be very similar to a homogeneous volume source, see also Appendix A.2.

The decomposition kinetics of the metal hydride phase system, see Sect. 2.5.2.2, in the metal melt is omitted in the present model. Thus, effects like the time-temperature dependence of the decomposition as well as the gas concentration already present in the melt are not taken into account.

7.2.5 Gas Diffusion

Diffusion phenomena are described by the diffusion equation, Eq. (4.7), where the boundary conditions at the fluid–gas interface trigger a gas current from the fluid to the gas bubbles. Generally, these boundary conditions are given by Sieverts' law, Eq. (4.9), describing the equilibrium between the gas pressure in the bubble and the gas dissolved in the fluid metal. The increase of the amount of gas in the bubbles leads to an increase of the bubble pressure which eventually leads to bubble expansion described by the NSEs, Eqs. (4.1) and (4.2).

[4]This is the situation during IFM.

7.2.6 Stabilization

A central point of the model is foam stabilization. Fluid dynamics leads to the development of transient cell walls but not to stable foam structures, see Sect. 4.2.3. Foam stabilization is realized by the introduction of a phenomenological local force, the disjoining pressure, see Sect. 4.3.2. In case of IFM, the disjoining pressure describes the effect of solid phase particles which get captured in the cell walls during foaming thereby inducing a stabilizing counter pressure Π, see Fig. 7.3.[5]

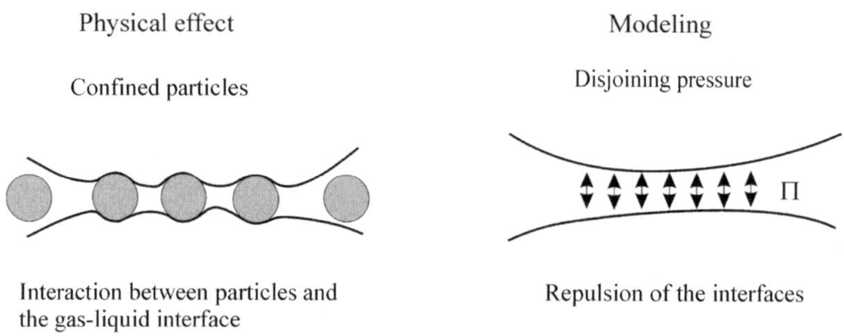

Figure 7.3: Foam stabilization. Physically, stabilization is the result of particle confinement. Numerically, the physical barrier effect is modeled by a disjoining pressure. The disjoining pressure Π is a function of the smallest distance to the neighboring interface.

The disjoining pressure has a finite range d_Π and increases linearly with decreasing wall thickness d.[6] We use the most simplest function which satisfies these requirements:

$$\Pi(d) = \begin{cases} c_\Pi \left(1 - \frac{d}{d_\Pi}\right) & \text{for} \quad d < d_\Pi \\ 0 & \phantom{\text{for}} \quad d \geq d_\Pi \end{cases} \tag{7.1}$$

where c_Π is a positive constant.

The function described by Eq. (7.1) is depicted in Fig. 7.4. The critical cell wall thickness d_{crit} is defined as the cell wall thickness at the onset of destabilization followed by rapid cell wall rupture.

7.2.7 Cell Wall Rupture and Coalescence

A further essential task is the description of cell wall rupture and bubble coalescence events. From experiments we know that thick cell walls are quite stable until they

[5]The evolution of a disjoining pressure by confined particles is discussed in detail in Chap. 6.
[6]The wall thickness is equivalent with the smallest interface distance.

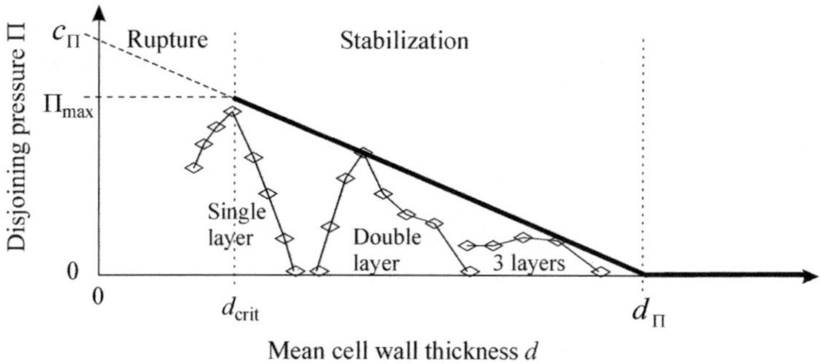

Figure 7.4: Disjoining pressure, Eq. (7.1), full line. Data points: Oscillating disjoining pressure produced by confined particles in 2D, see Fig. 6.7. The critical thickness d_{crit} designates the onset of destabilization followed by cell wall rupture.

reach a critical thickness d_{crit}. On the other hand, cell wall rupturing proceeds in milliseconds [15, 16]. The transition from a stable to an instable cell wall and eventually the process of rupture is very complex and far beyond the modeling capacities.

Analogous to stabilization, we use a phenomenological approach to model cell wall rupture. This model is based on the existence of a critical cell wall thickness d_{crit}, see Fig. 7.4. If the cell wall undergoes d_{crit} part of the wall is locally completely removed, see Fig. 7.5

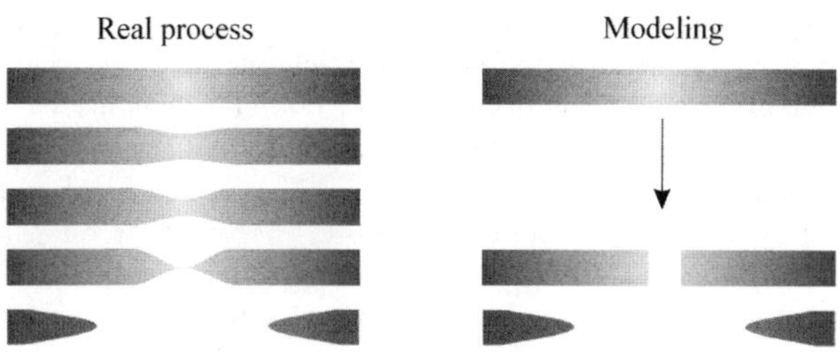

Figure 7.5: Cell wall rupture. Local thinning precedes rupture in metal foams. This process is simplified in the model by a local removal of the wall.

Cell wall rupture is assumed to occur instantaneously at the critical thickness and the exact thinning process is not described. This drastic simplification is reasonable due to the fact that thinning is a local process. The outcome of this simple rupture description is very similar to that of the actual process. A more detailed description would not lead to much more insight. The local break up of the cell wall leads to bubble coalescence, i. e. the two bubbles merge and their gas contents are added.

7.2.8 Temperature

The current model does not contain thermal diffusion, i. e. foaming takes place at constant temperature. Consequently, the material parameters, viscosity, surface tension and hydrogen solubility, are constant.

7.3 Lattice Boltzmann Approach

Conventional methods of CFD are based on a discretization and numerical solution of the NSEs e. g. by finite differences, finite volume or finite element methods [141,206]. The lattice Boltzmann method (LBM) is an alternative to ordinary CFD approaches. It is based on a kinetic picture with the famous Boltzmann equation as foundation. Pseudo-particles are moving and interacting according to a set of rules in a fictitious world obtained by discretization. This fictitious world is called a cellular automaton (CA) [240]. The hydrodynamic behavior emerges in a natural way from the CA by incorporating the correct physics.

The lattice Boltzmann method can be seen as a mesoscopic approach placed somewhere in between macroscopic CFD and microscopic molecular dynamics. It has the potential to fill the gap where fluid dynamics breaks down and molecular dynamics is not yet applicable for lack of computing power. The potential of LBM is strongly coupled with this dual fluid-particle nature [43, 44, 117, 219]. On the one hand, LBM is a solver for the macroscopic NSEs. On the other hand, there are a lot of benefits which are directly related to the particle picture, e. g. local conservation of mass and momentum, an easy formulation of boundary conditions, etc.

There are different reasons why the LBM seems to be especially appropriate for the present foaming problem. Firstly, the ability of the LBM to treat grossly irregular boundary conditions in an efficient way.[7] Secondly, to allow explicit mass conservation. The latter property is especially important since the fluid–gas interface is not only complex but also strongly evolving with time.

The application of the LBM for foam formation simulations implies the existence of an algorithm to deal with free surface problems. Before we describe the free surface algorithm (Chap. 8) and the LBM for foam evolution (Chap. 9), a short introduction into the main principles of CA and the LB approach is given.

[7] Flows through porous media are one of the most successful applications of LBM, see e. g. [20,195]. In addition, multi-phase fluid motion [222] and phase transitions [140] represent important fields of application.

7.3.1 Cellular Automata

Our approach to model foams is based on the LBM which belongs to the class of cellular automata (CA).[8] It is therefore instructive to start the theoretical part with a short introduction into the fundamental principles of CA. These simple principles are preserved in lattice Boltzmann approaches and form their foundation. Cellular automata represent a physical system in an idealized way where space and time are discrete, i. e. a fully discrete universe made up of identical cells.[9] The main characteristics of a CA are the following [43]:

- Regular lattice of cells in m dimensions.
- Each cell at lattice site $\mathbf{x}$ is characterized by a set of b Boolean variables ϕ_i defining the local state $\Phi(\mathbf{x}, t)$ of each cell: $\Phi(\mathbf{x}, t) = \{\phi_1, \ldots, \phi_b\}$.
- The evolution rule is the same for all cells and updating of the cells occurs simultaneously in discrete time steps.
- The evolution rule is a function of the state of the neighborhood. The neighborhood normally comprises the central cell and adjacent cells (e.g., see Fig. 7.6).

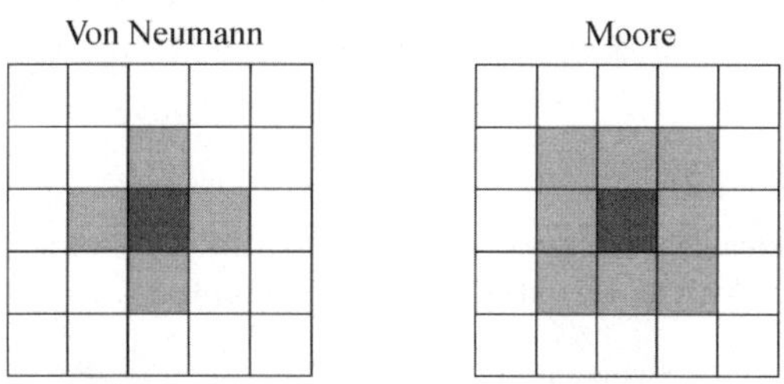

Figure 7.6: Von Neumann and Moore neighborhoods. The *shaded region* indicates the cells whose state determines the state of the central cell (*dark gray*) during updating.

The essential feature of CA is their ability to model complex systems [83]. Although built up from simple interacting components a very complex collective behavior can emerge during time evolution. The global behavior obeys new laws which can not be simply extrapolated from the individual behavior. That is, the whole is more than the sum of the components. This is the reason why the CA approach

[8]Cellular automata were invented by John von Neumann in the late 1940s. His aim was to build a machine with the capability of self-reproduction, the main feature of biological systems.

[9]CA cells must not be mixed up with foam cells.

is especially interesting to simulate complex and non-equilibrium phenomena. Cellular automata models are not only toy systems whose behavior is similar to real dynamical systems. They can represent physical systems and are able to produce macroscopic results. The source of this property is founded in statistical mechanics where the macroscopic behavior of many systems is quite independent from its microscopic one. Only symmetries and conservation laws are transferred from the microscopic to the macroscopic world. A CA model can be seen as a fictitious universe which has its own microscopic reality but has the same macroscopic behavior as the real system we are interested in [43].

A CA is defined by its states and its evolution rules. Time evolution of a CA is normally characterized by two steps, a **streaming step** followed by a **collision (= relaxation) step**. During streaming, information of each cell is transfered to the neighborhood. The collision step comprises rules how the incoming information is processed in order to model specific microscopic interactions, e. g. collision processes or reactions.

7.3.1.1 Lattice Gas Automata

Although we will not go into details of CA it is instructive to have a closer look at lattice gas automata since the main principles of the LBM are already existing at this level. Lattice gas automata describe the dynamics of point particles moving and colliding in a discrete space-time universe. A lattice gas model consists of the following elements:

- A regular lattice of cells.
- A set of Boolean variables ϕ_i, $i \in \{1, ..., b\}$, the state of a cell, describing the population of b given velocities $\mathbf{e}_i$ ($\phi_i = 0$: empty, $\phi_i = 1$: occupied).

The dynamics of a lattice gas model is divided into two steps:

1. Streaming Advection from cell to cell according to the velocities $\mathbf{e}_i$.
2. Collision Redistribution of the populations at each cell during the collision step while preserving mass and momentum.

Figure 7.7 shows the streaming and collision of a so-called HPP gas model [43] with $\mathbf{e}_1 = (1,0), \mathbf{e}_2 = (0,1), \mathbf{e}_3 = (-1,0), \mathbf{e}_4 = (0,-1)$ (von Neumann neighborhood). The collision rule for the HPP gas is expressed as (HPP rule)

$$\{1010\} \longrightarrow \{0101\} \qquad \{0101\} \longrightarrow \{1010\}. \tag{7.2}$$

All other configurations are unchanged.

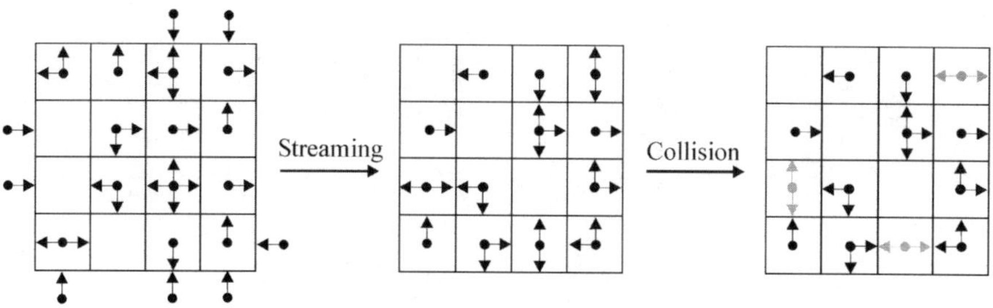

Figure 7.7: HPP gas. Initial configuration (*left*), streaming (*middle*) and HPP collision (*right*). Gray arrows indicate an alteration due to collision.

Based on these models it was for the first time realized that lattice gas models are in fact able to describe macroscopic hydrodynamic problems.[10] Figure 7.8 shows the development and propagation of a HPP sound wave. In addition, two other prominent CA examples are depicted [43].

The main advantages of lattice gas techniques are besides simple numerical codes their stability, easy introduction of boundary conditions and intrinsic parallel structure allowing high performance computing [155]. It can be shown that lattice gas models with an appropriate choice of the lattice symmetry in fact represent numerical solutions of the NSEs (FHP model by Frisch, Hasslacher and Pomeau [74]). The advantages of CA approaches become apparent when complex boundary conditions are present. In this case, the microscopic interpretation of the dynamics allows to treat boundary conditions in a much more natural way than the continuous description which is based on differential equations.[11]

However, lattice gas models suffer from some drawbacks: statistical noise, non-Galilean invariance, a velocity dependent pressure and spurious invariants. These difficulties are intimately related to the Boolean variables of the lattice gas with its underlying Fermi-Dirac or Bose-Einstein statistics. Particularly, the statistical noise requires time-space averaging procedures to extract macroscopic quantities like the density or the velocity. This intrinsic property of LGA is the reason why they were not able to surpass conventional numerical methods of hydrodynamics. Nevertheless, LGA have been very successful where traditional computing techniques are not or only with strong difficulties applicable, e. g. flows in porous media, immiscible flows [104], micro emulsions, etc. [43].

[10]However, HPP gases suffer from a lack of isotropy and spurious invariants.

[11]We explicitly make use of this advantage for the free boundary conditions, see Chap. 8.

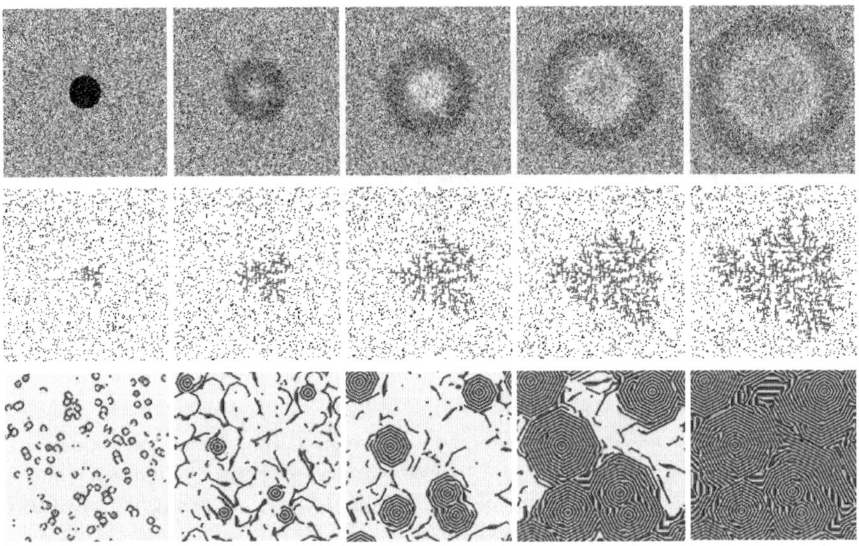

Figure 7.8: Famous examples for CA. *Top*: Sound wave (Lattice gas automaton: HPP gas), *Middle*: Diffusion limited aggregation, *Bottom*: Excitable media.

7.3.2 Lattice Boltzmann Method

Historically, the lattice Boltzmann method developed from lattice gas automata. McNamara and Zanetti [178] were the first who extended the Boolean dynamics of the automaton to real numbers representing the probability for a cell to have a given state. The philosophy behind this procedure is, that it is more efficient to average the microdynamics before than after simulation. That is, the discrete nature of the fluid particles vanishes on the macroscopic level of observation. The LBM is characterized by a much higher numerical efficiency than the Boolean dynamics. In addition, lattice Boltzmann methods maintain the intuitive microscopic level of interpretation[12] belonging to CA. These properties make LBMs promising approaches to model complex physical systems.

In the following, the foundation of the LBM is described. We do not follow the historical way but the more profound derivation by He and Luo [106,107] much later. He and Luo were the first who showed that the LBM is not only a CA which gives – more or less by accident – right hydrodynamic results, but a subtle discretization of the Boltzmann equation. Thus, the LBM is founded on a sound physical framework.

[12]That is, particles moving from one lattice site to the other.

7.3.2.1 Boltzmann Equation

The Boltzmann equation[13] is the foundation of the kinetic theory, which is the branch of statistical physics dealing with the dynamics of non-equilibrium processes and their relaxation to thermodynamic equilibrium. The governing equations of fluid dynamics, which we focus on, emerge from the kinetic theory in the mean field picture by a perturbative treatment of the kinetic equations.[14]

The Boltzmann equation describes the temporal evolution of single particle distribution functions $f(\mathbf{x}, \boldsymbol{\xi}, t)$ where f is defined such that $f(\mathbf{x}, \boldsymbol{\xi}, t)\, d\mathbf{x}\, d\boldsymbol{\xi}$ is the number of particles or molecules at time t positioned between $\mathbf{x}$ and $\mathbf{x} + d\mathbf{x}$ with velocities between $\boldsymbol{\xi}$ and $\boldsymbol{\xi} + d\boldsymbol{\xi}$: **Boltzmann equation**

$$\frac{\partial f}{\partial t} + \boldsymbol{\xi} \cdot \boldsymbol{\nabla} f + \mathbf{F} \cdot \frac{\partial f}{\partial \boldsymbol{\xi}} = S. \tag{7.3}$$

The left hand side describes reversible Newtonian single-particle dynamics, i. e. streaming motion of the molecules along trajectories associated with the force field $\mathbf{F}$. The collision operator S on the right hand side is the sum of all intermolecular interactions. These collisions take molecules in or out the streaming trajectory.

Relaxation Time Approximation The basis for the most widely used lattice Boltzmann equation (LBE) is the Boltzmann equation in **BGK approximation**,

$$\frac{\partial f}{\partial t} + \boldsymbol{\xi} \cdot \boldsymbol{\nabla} f + \mathbf{F} \cdot \frac{\partial f}{\partial \boldsymbol{\xi}} = -\frac{1}{\tau} [f - f^{\mathrm{eq}}], \tag{7.4}$$

where the collision operator S is replaced by a simplified collision operator, the single relaxation time approximation due to Bhatnagar, Gross, and Krook (BGK model) [22] with the relaxation time τ. The relaxation time characterizes the time necessary to reach the equilibrium distribution by collisions.

The equilibrium distribution function f^{eq} is the **Maxwell-Boltzmann equilibrium distribution function** [40],

$$f^{\mathrm{eq}}(\boldsymbol{\xi}) = \rho \left(\frac{m}{2\pi k_{\mathrm{B}} T}\right)^{3/2} \exp\left[-\frac{m\,(\boldsymbol{\xi} - \mathbf{v})^2}{2 k_{\mathrm{B}} T}\right], \tag{7.5}$$

with the Boltzmann constant k_{B}, the particle mass m and the temperature T. Mass density ρ and momentum density $\rho\mathbf{v}$ follow from the first moments of the distribution function,

$$\rho = \int f d\boldsymbol{\xi}, \qquad \rho\mathbf{v} = \int \boldsymbol{\xi} f d\boldsymbol{\xi}. \tag{7.6}$$

[13] The Boltzmann equation was established by Ludwig Boltzmann in 1872. He committed suicide in 1906.

[14] For details, see e. g. [43].

Discrete Velocity Model It was first shown by He and Luo [106, 107] that the particle velocity space $\boldsymbol{\xi}$ may be reduced to a small set of discrete velocities $\{\boldsymbol{\xi}_i | i = 1, \ldots, b\}$ while preserving the hydrodynamic moments up to a certain order in $\boldsymbol{\xi}$. By a proper discretization[15] of the velocity space $\boldsymbol{\xi}$, Eq. (7.4) is reduced to a **discrete Boltzmann equation**,

$$\frac{\partial f_i}{\partial t} + \boldsymbol{\xi_i} \cdot \boldsymbol{\nabla} f_i = -\frac{1}{\tau} \left[f_i - f_i^{\mathrm{eq}} \right], \tag{7.7}$$

where $f_i(\mathbf{x}, t) \equiv f(\mathbf{x}, \boldsymbol{\xi}_i, t)$ and $f_i^{\mathrm{eq}}(\mathbf{x}, t) \equiv f^{\mathrm{eq}}(\mathbf{x}, \boldsymbol{\xi}_i, t)$ denote the distribution function and the equilibrium distribution function of the ith discrete velocity $\boldsymbol{\xi}_i$, respectively. The discretization of the force term of the continuous Boltzmann equation, $\mathbf{F} \cdot \partial f / \partial \boldsymbol{\xi}$, is not straight forward since the velocities $\boldsymbol{\xi}_i$ are constant, i. e. the derivative can not be simply computed.[16]

7.3.2.2 Lattice Boltzmann Equation

The lattice Boltzmann equation follows from the discrete Boltzmann equation, Eq. (7.7), by space-time discretization. In the following, the equations are formulated in dimensionless representation. The discrete velocity set $\{\xi_i\}$ is thus replaced by the dimensionless discrete velocity set $\mathbf{e}_i$ with $\{\xi_i\} = c \cdot \{\mathbf{e}_i\}$ where c is an arbitrary constant with dimension of a velocity.[17]

The discretization in space and time has to be coherent with the velocity discretization, i. e. each velocity vector $\mathbf{e}_i$ has to coincide with the displacement vector from a cell to one of its neighbors. The space-time discretized Boltzmann equation reads [19]:

$$f_i(\mathbf{x} + \mathbf{e}_i, t + 1) - f_i(\mathbf{x}, t) = -\frac{1}{\tau} \left[f_i(\mathbf{x}, t) - f_i^{\mathrm{eq}}(\mathbf{x}, t) \right]. \tag{7.8}$$

Eq. (7.8) is the famous **lattice Boltzmann equation (LBE)** in BGK approximation [40, 110, 178, 200]. The LBE is a discretized version of the Boltzmann equation where the continuous phase space is reduced to only a few discrete points. Analogous to Eq. (7.6), the macroscopic density ρ and momentum $\rho\mathbf{v}$ at a cell are the 0th and 1th moments of the distribution functions,

$$\rho = \sum_{i=0}^{b} f_i, \qquad \rho\,\mathbf{v} = \sum_{i=0}^{b} f_i\,\mathbf{e}_i. \tag{7.9}$$

[15] It is instructive to notice that the discretization is not unique and strongly depends on the dimension of the problem. This ambiguousness leads to the existence of several lattice Boltzmann models denoted by DnQm, see Sect. 7.3.2.3.

[16] However, there are different approaches to include a force term, see Sect. 7.3.2.4.

[17] The physical quantity follows by multiplying with the relevant time, length, mass and amount of substance scale denoted by Δt, Δx, Δm and Δn, respectively. See also Appendix C.

Analogous to the lattice gases in Sect. 7.3.1.1, the LBE is solved in two steps, streaming and collision:

1. **Streaming**
$$f_i^{\mathrm{in}}(\mathbf{x},t) = f_i^{\mathrm{out}}(\mathbf{x}-\mathbf{e}_i, t-1) \tag{7.10}$$

2. **Collision**
$$f_i^{\mathrm{out}}(\mathbf{x},t) = f_i^{\mathrm{in}}(\mathbf{x},t) - \frac{1}{\tau}\left[f_i^{\mathrm{in}}(\mathbf{x},t) - f_i^{\mathrm{eq}}(\mathbf{x},t)\right] \tag{7.11}$$

The incoming and outgoing distribution functions, i. e. before and after collision, are denoted with f_i^{in} and f_i^{out}, respectively. During streaming, Eq. (7.10), all distribution functions but f_0 are advected to their neighbor cells defined by their velocity. After advection, the particle distribution functions approach their equilibrium distributions due to a collision step, Eq. (7.11).

The **equation of state** is that of an ideal gas where the pressure p and the density ρ are correlated [219],

$$p = c_{\mathrm{s}}^2\,\rho, \tag{7.12}$$

where c_{s} is the **speed of sound**.

The **viscosity** ν is determined by the relaxation time τ [219],

$$\nu = c_{\mathrm{s}}^2(\tau - 0.5). \tag{7.13}$$

7.3.2.3 DnQm Models

A lattice Boltzmann model is defined by its discrete velocity set $\{\mathbf{e}_i\}$ and the form of the equilibrium distribution function. Qian et al. [200] provide a whole family of solutions, designated as DnQm, where n stands for the spatial dimension and m for the number of different velocity vectors, see Fig. 7.9. For the fluid dynamic problem in 2D and 3D, we use the D2Q9 and D3Q19 model, respectively [106]. The D2Q4 model is employed for 2D gas diffusion [238, 239].[18]

The form of the equilibrium distribution functions must be chosen in such a way that fluid mass and momentum are conserved while the resulting continuum equations describe the hydrodynamics of the fluid. In addition, isotropy and Galilean

[18] The gas diffusion problem is a scalar equation whereas the NSEs are vector equations. This is the reason why a simpler model can be used for the diffusion problem. The D2Q4 model is not sufficient for the hydrodynamic problem.

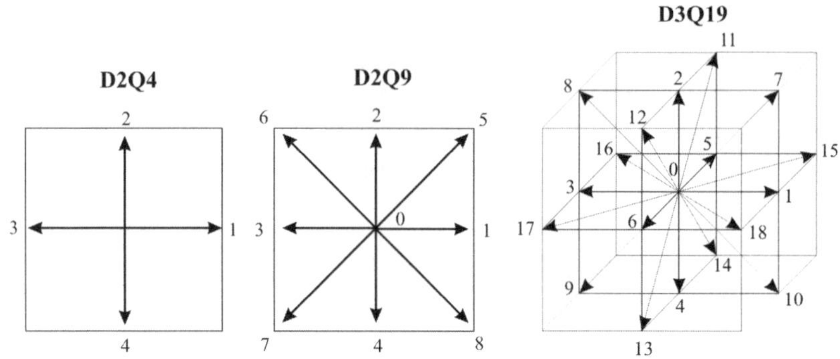

Figure 7.9: Velocity sets for different lattice Boltzmann models: D2Q4, D2Q9 and D3Q19. Rest particles are defined in the D2Q9 and D3Q19 models.

invariance have to be fulfilled. The equilibrium functions for fluid dynamics, f_i^{eq}, follow from an expansion of the Maxwell-Boltzmann equilibrium distribution function, Eq. (7.5), in a Taylor series of $\mathbf{v}$ up to second order [106]:

$$f_i^{\mathrm{eq}}(\rho(\mathbf{x},t),\mathbf{v}(\mathbf{x},t) = f_i^{\mathrm{eq}}(\mathbf{x},t) = w_i\,\rho\left[1+3\,(\mathbf{e}_i\cdot\mathbf{v})+\frac{9}{2}\,(\mathbf{e}_i\cdot\mathbf{v})^2-\frac{3}{2}\,\mathbf{v}^2\right]. \qquad (7.14)$$

The coefficients w_i depend on the discrete velocity set $\{\mathbf{e}_i\}$, see Table 7.1 [219].

Table 7.1: Parameters for some DnQm BGK lattices.

	c_s^2	$\frac{1}{2}\mathbf{e}_i^2$	w_i
D2Q4	1/2	1/2	1/4
D2Q9	1/3	0	4/9
		1/2	1/9
		1	1/36
D3Q19	1/3	0	12/36
		1/2	2/36
		1	1/36

7.3.2.4 LBM with Gravity

The incorporation of the body force term, $\mathbf{F} \cdot \partial f / \partial \boldsymbol{\xi}$, in the LBE is not straight forward since the velocity of all the 'particles' is constant in the LBM. Consequently, we can not employ a term with exact the same form but must instead introduce a fluid momentum changing term.

The first approaches to introduce a body force term were solely phenomenological [202]. A survey of strategies to introduce gravity is given by Buick and Greated [36]. Two popular approaches are as follows:

- Addition of a term to the collision function that modifies the distribution function,

$$f_i(\mathbf{x} + \mathbf{e}_i, t + 1) - f_i(\mathbf{x}, t) = -\frac{1}{\tau} \left[f_i(\mathbf{x}, t) - f_i^{\mathrm{eq}}(\mathbf{x}, t) \right] + a_i \left(\mathbf{F} \cdot \mathbf{e}_i \right), \tag{7.15}$$

  where the a_i are suitable weighting factors.

- Alteration of the velocity in the equilibrium distribution function,

$$f_i^{\mathrm{eq}}(\rho, \mathbf{v}) \quad \longmapsto \quad f_i^{\mathrm{eq}}(\rho, \mathbf{v} + b_i \left(\mathbf{F} \cdot \mathbf{e}_i \right)), \tag{7.16}$$

  where the b_i are suitable weighting factors.

Buick and Greated [36] show that neither of these two approaches is exact. Only a combination of both leads to the NSE without additional term.

A more profound derivation by Luo [174] considers the expansion of the moments of the body force. Gravity is introduced into the D2Q9 or D3Q19 LB approach analogous to Eq.(7.15) but with additional terms proportional to the velocity:

$$f_i(\mathbf{x}+\mathbf{e}_i, t+1) - f_i(\mathbf{x}, t) = -\frac{1}{\tau} \left[f_i(\mathbf{x}, t) - f_i^{\mathrm{eq}}(\mathbf{x}, t) \right] + 3 w_i \, \rho \left[(\mathbf{e}_i - \mathbf{v}) + 3 (\mathbf{e}_i \cdot \mathbf{v}) \, \mathbf{e}_i \right] \cdot \mathbf{g}, \tag{7.17}$$

where $\mathbf{g}$ is the gravitational acceleration. We use Eq. (7.17) to incorporate gravity.

The outcome of the additional term in Eq. (7.17) is that the definition of the fluid momentum has to be renewed. The fluid momentum $\rho \tilde{\mathbf{v}}$ is defined to be the average of the momentum before and after the collision,

$$\rho \tilde{\mathbf{v}} = \rho \mathbf{v} + \frac{1}{2} \rho \mathbf{g}. \tag{7.18}$$

8 Lattice Boltzmann Model for Free Surface Flow

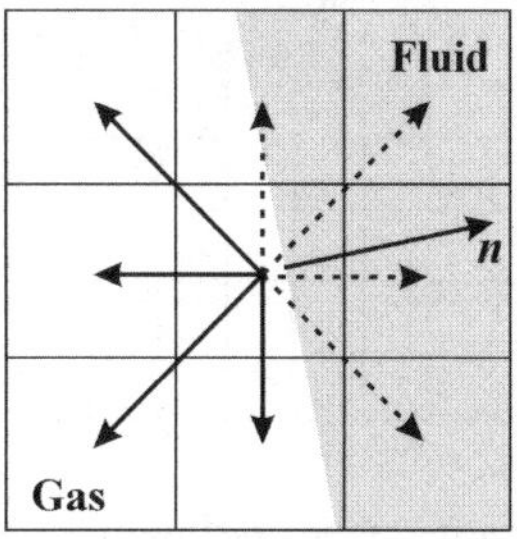

The main difficulty to apply a lattice Boltzmann approach on foam evolution problems is the correct treatment of the free surface flow [45, 154, 155, 160, 161, 162, 163, 224]. The two fundamental challenges consist in the advection of the free interface and the fulfillment of the physical pressure boundary conditions. The isotropic advection of an interface located within a non-isotropic lattice without changes of the total mass or broadening is rather difficult. In addition, to guarantee the correct boundary conditions in the framework of the LBM, where we deal with a set of particle distribution functions, is a challenge.[1] In the following, the basic algorithms applicable for 2D and 3D are described.

[1]At the beginning of this work a theoretical basis for the treatment of free surface flows in the framework of a lattice Boltzmann model was not available in literature. Now, there exists an alternative technique to deal with free surface problems with a LBM (see Ginzburg and Steiner [86, 87]). Ginzburg and Steiner propose a volume tracking LB method which can be seen as a modified immiscible lattice Boltzmann model [98, 99] with two components, fluid and vacuum. Surface tension, bubbles or bubble coalescences are not treated.

8.1 Free Surface and Fluid Advection

8.1.1 Representation of the Free Surface

The description of the fluid–gas interface is similar to that of volume of fluid methods. An additional variable, the volume fraction of fluid ϵ, defined as the portion of the area of the cell filled with fluid, is assigned to each interface cell. The representation of fluid–gas interfaces is depicted in Fig. 8.1.

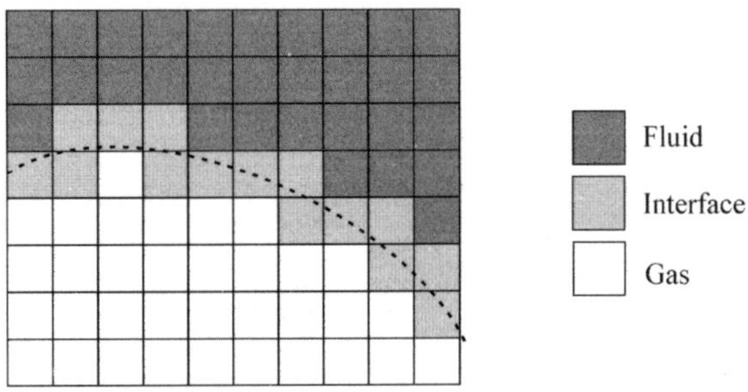

Figure 8.1: Representation of a free fluid–gas interface by interface cells. The real interface (dashed line) is captured by assigning to the interface cells their fluid fraction.

Gas cells are separated from fluid cells by a layer of interface cells. These interface cells form a completely closed boundary in the sense that no distribution function is directly advected from fluid to gas cells and vice versa. This is a crucial point to assure mass conservation since mass coming from the fluid or mass transferred to the fluid always passes through interface cells where the total mass is balanced. Hence, global conservation laws are fulfilled if mass and momentum conservation is ensured for interface cells. The used cell types and their state variables and possible state transformations are listed in Table 8.1.

8.1.2 Mass Exchange

Per definition, the volume fraction ϵ of fluid and gas cells is 1 and 0, respectively. The fluid mass content of a cell is denoted by $M = M(\mathbf{x}, t)$. For a gas cell the fluid mass content M is zero whereas that of a fluid cell is given by its density ρ and the cell volume ΔV:

$$M(\mathbf{x}, t) = \rho(\mathbf{x}, t) \cdot \Delta V, \qquad \mathbf{x} \in F. \tag{8.1}$$

Table 8.1: Cell types: state variables and possible state transformations. Wall cells, which describe rigid boundaries, are designated as W. •: state variable defined. –: state variable not defined.

Cell type	f_i	Fluid fraction (ϵ)	Gas pressure (p^{G})	Change of state
Fluid F	•	–	–	$\rightarrow$ I
Gas G	–	–	•	$\rightarrow$ I
Interface I	•	•	•	$\rightarrow$ G , $\rightarrow$ F
Wall W	•	–	–	–

Fluid cells gain and lose mass due to streaming of the f_i. For fluid cells M and ρ are equivalent. The situation is more complicated if interface cells are considered. Cells are interface cells if the macroscopic interface runs through it. That is, only part of the cell is covered with fluid. In this case, M and ρ are not equivalent and we have to account for the partially filled state by introducing a second parameter, the fluid volume fraction $\epsilon = \epsilon(\mathbf{x}, t)$. The fluid mass content M, the fluid volume fraction ϵ and the density ρ are related by

$$M(\mathbf{x}, t) = \rho(\mathbf{x}, t) \cdot \epsilon(\mathbf{x}, t) \cdot \Delta V, \qquad \mathbf{x} \in \mathrm{I}. \tag{8.2}$$

All cells but wall cells are able to change their state. It is important to note that direct state changes from fluid to gas and vice versa are not possible. Hence, fluid and gas cells are only allowed to transform into interface cells whereas interface cells can be transformed into both gas and fluid cells.

A fluid cell is transformed into an interface cell if a cell of its neighborhood is transformed into a gas cell. At the moment of transformation the fluid cell contains a certain amount of fluid mass M which is recorded. During further development the interface cell can gain mass from or lose mass to the neighboring cells. These mass currents are calculated and lead to a temporal change of M. If M drops below zero, the interface cell is transformed into a gas cell. It is important to pronounce that mass and density are completely decoupled for interface cells. While the density of the interface cells is given by the pressure boundary conditions and fluid dynamics, M is determined by the mass exchange with the neighboring fluid and interface cells.

The mass exchange $\Delta M_i(\mathbf{x}, t)$ between an interface cell at lattice site $\mathbf{x}$ and its neighbor in $\mathbf{e}_i$-direction at $\mathbf{x} + \mathbf{e}_i$ is calculated from

$$\Delta M_i(\mathbf{x}, t) = A_{\mathrm{eff}} \cdot \left(f_{\bar{\imath}}^{\mathrm{out}}(\mathbf{x} + \mathbf{e}_i, t) - f_i^{\mathrm{out}}(\mathbf{x}, t) \right), \qquad \mathbf{x} \in I, \tag{8.3}$$

where $\bar{\imath}$ denotes the inverse direction: $\mathbf{e}_{\bar{\imath}} = -\mathbf{e}_i$. The effective area A_{eff}, $0 \leq A_{\mathrm{eff}} \leq 1$, can be geometrically interpreted as the fraction of the cell boundary where mass

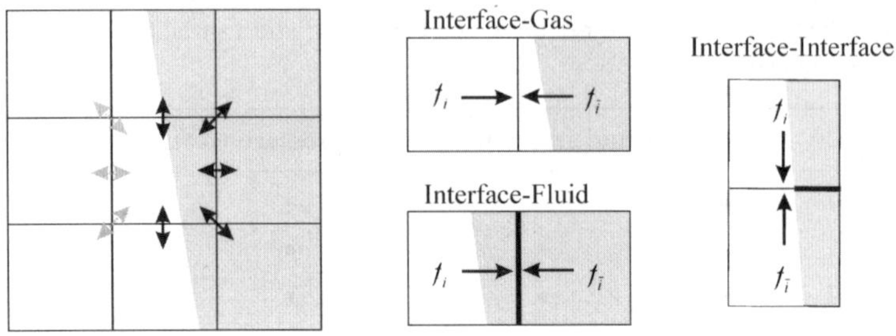

Figure 8.2: Mass exchange at interface cells. There are three different cases depending on whether the neighboring cell is a gas, fluid or interface cell. The effective transfer areas A_{eff} are indicated by *bold lines*.

is actually transferred, see Fig. 8.2. There is no mass transfer between gas cells and interface cells since there is only gas at the common cell boundary and thus $A_{\text{eff}} = 0$. The interchange between an interface cell and a fluid cell is the same as that of two fluid cells since the common cell boundary is completely covered with fluid and therefore $A_{\text{eff}} = 1$. Otherwise, a straight interface line could not move with constant velocity through the lattice. What remains is the effective area for two interface cells. In principle, for cells in coordinate direction this area can be estimated by a geometric reconstruction of the actual interface. In this geometric picture the problem arises how the interchange between two diagonal cells has to be treated since two diagonal cells have no cell boundary in common. Nevertheless, the effective area for two interface cells is always estimated by the mean value of their fluid contents. This approach has only a geometric interpretation for some cases. The following equations summarize the phenomenological algorithm for the mass exchange of an interface cell at lattice site $\mathbf{x} \in \mathrm{I}$:

$$\Delta M_i(\mathbf{x},t) = \begin{cases} 0 \\ f_{\bar{i}}^{\text{out}}(\mathbf{x}+\mathbf{e}_i,t) - f_i^{\text{out}}(\mathbf{x},t) \\ \frac{1}{2}(\epsilon(\mathbf{x},t)+\epsilon(\mathbf{x}+\mathbf{e}_i,t))(f_{\bar{i}}^{\text{out}}(\mathbf{x}+\mathbf{e}_i,t) - f_i^{\text{out}}(\mathbf{x},t)) \end{cases} \quad \mathbf{x}+\mathbf{e}_i \in \begin{cases} \mathrm{G} \\ \mathrm{F} \\ \mathrm{I} \end{cases} \tag{8.4}$$

It is crucial to note that this algorithm explicitly conserves mass:

$$\Delta M_{\bar{i}}(\mathbf{x}+\mathbf{e_i},t) = -\Delta M_i(\mathbf{x},t), \qquad x \in I. \tag{8.5}$$

Mass which a particular cell receives from a neighboring cell is automatically lost there and vice versa. The temporal evolution of the mass content of an interface cell is thus given by

$$M(\mathbf{x},t+1) = M(\mathbf{x},t) + \sum_{i=1}^{b} \Delta M_i(\mathbf{x},t), \qquad x \in I. \tag{8.6}$$

An interface cell is transformed into a gas or fluid cell if $M < 0$ or $M > \rho\,\Delta V$, respectively. At the same moment, new interface cells emerge in order to guarantee the continuity of the interface. The initial distribution functions of these new interface cells are extrapolated from the cells in normal direction towards the fluid.[2]

8.2 Interface Boundary Conditions

8.2.1 Missing Distribution Functions

Interface cells separate gas cells from fluid cells. After streaming, only distribution functions from fluid and interface cells are given. Distribution functions arriving from gas cells are not defined (see Fig. 8.3, left) and have to be reconstructed in such a way that the pressure boundary conditions are fulfilled.

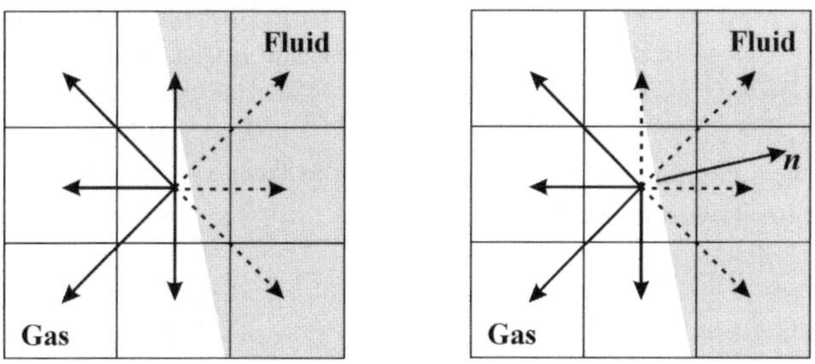

Figure 8.3: Missing distribution functions at interface cells after streaming. *Left*: Undefined distribution functions after streaming (*broken lines*). Right: Set of distribution functions with $\mathbf{n} \cdot \mathbf{e_i} \geq 0$ (*broken lines*), $\mathbf{n}$: normal vector.

The macroscopic boundary conditions, see Eqs. (4.5) and (4.6), express that the force exerted by the gas on the fluid balances the force exerted by the fluid on the gas. The force acting on a surface element results from the momentum flux through it. Conventional CFD determines the momentum flux by means of computation of the viscous tensor $\sigma_{\alpha\beta} = \eta\left(\frac{\partial v_\alpha}{\partial x_\beta} + \frac{\partial v_\beta}{\partial x_\alpha}\right)$. Consequently, CFD has to apply procedures to extract gradient information of the velocity field and space derivatives have to be known explicitly.

[2]In order to prevent broadening of the interface special treatment for some situations is necessary. For details see [223].

In the following, a reconstruction procedure for the missing distribution functions is presented where the kinetic nature of the LBM is explicitly exploited. Our approach to fulfill the pressure boundary conditions is based on a momentum-exchange method which was first applied by Ladd [165, 166, 167] to compute the fluid force on a sphere in suspension flow. Mei et al. [179] have successfully applied a similar approach to calculate forces for problems where curved geometries are involved. The treatment of free interface flows in the momentum-exchange method is not known in literature and presented in the following. Our approach is very similar to that of Chen et al. [38] who present a method to fulfill hydrodynamic boundary conditions at curved solid surfaces.

8.2.2 Force Equilibrium

The underlying philosophy of the reconstruction step is that we know the gas pressure and from there the force exerted by the gas on the fluid. That is, the boundary conditions are fulfilled if the force exerted by the fluid balances the gas force. The total force is generated by the particles crossing the interface, i. e. the particle currents from the fluid to the gas and vice versa. Only fluid particles with appropriate velocity, $\mathbf{n} \cdot \mathbf{e}_i < 0$ (Fig. 8.3, right), will traverse the interface during the advection step.[3] The same is true for particles originating from the gas. In this case, the velocity of the particles points inwards, $\mathbf{n} \cdot \mathbf{e}_i \geq 0$. Thus, the total force $\boldsymbol{F}$ exerted by the fluid on a surface element $\boldsymbol{A}(\mathbf{x}) = \mathbf{n} {\cdot} A(\mathbf{x})$ results from the momentum transported by the particles streaming through this element during one time step Δt, see Fig. 8.4:[4]

$$\begin{aligned} F_\alpha &= -n_\beta\, A(\mathbf{x}) \sum_{i,\mathbf{n}\cdot\mathbf{e}_i<0} f_i^{\rm out}(\mathbf{x},t)\,(e_{i,\alpha}-v_\alpha)(e_{i,\beta}-v_\beta) \\ &\quad -n_\beta\, A(\mathbf{x}) \sum_{i,\mathbf{n}\cdot\mathbf{e}_i\geq 0} f_i^{\rm out}(\mathbf{x}-\mathbf{e}_i,t)\,(e_{i,\alpha}-v_\alpha)(e_{i,\beta}-v_\beta). \end{aligned} \tag{8.7}$$

In Eq. (8.7), the pure advection term is eliminated by subtracting the interface velocity $\mathbf{v}$ from the particle velocities. At the interface, $b/2$ of the b velocity vectors point from gas to fluid while the others point from fluid to gas. The first sum contains all known distribution functions coming from the fluid going into the gas. The second sum contains all unknown distribution functions coming from the gas and going into the fluid.[5] The unknown incoming distribution functions ($\mathbf{n} \cdot \mathbf{e_i} \geq 0$) have to be chosen in such a way that the pressure boundary conditions are satisfied, i. e. the force $\boldsymbol{F}$ must balance the gas force.

[3] The normal vector is calculated by the harmonic average of the vectors of the neighborhood weighted by the volume fractions, see [223].

[4] $f_i^{\rm out}$ denotes the particle distribution functions after collision and before advection.

[5] There are also distribution functions with $\mathbf{n} \cdot \mathbf{e}_i \geq 0$ which are in fact known. Nevertheless, these functions are treated as unknown.

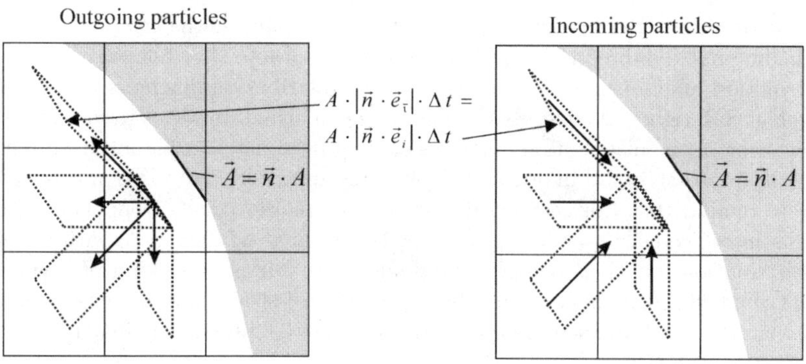

Figure 8.4: Momentum currents at interface cells. The momentum transferred to the interface area A is proportional to the number of particles which flow through it during one time step. *Left*: Outgoing momentum. *Right*: Incoming momentum.

8.2.3 Reconstruction of Unknown Distribution Functions

The symmetry between known and unknown distribution functions, i. e. if f_i is known $f_{\bar{\imath}}$ is unknown, is essential to fulfill the boundary conditions. We demand force balance for opposite lattice directions. In addition, we make use of the fact that the forces exerted by the gas are known and are given by the gas pressure and the velocity at the interface. Hence, the missing distribution functions are reconstructed as:

$$f_i^{\mathrm{out}}(\mathbf{x} - \mathbf{e}_i, t) = f_i^{\mathrm{eq}}(\rho^{\mathrm{G}}, \mathbf{v}) + f_{\bar{\imath}}^{\mathrm{eq}}(\rho^{\mathrm{G}}, \mathbf{v}) - f_{\bar{\imath}}^{\mathrm{out}}(\mathbf{x}, t), \qquad \forall i\colon\ \mathbf{n} \cdot \mathbf{e}_i \geq 0 \quad (8.8)$$

with the gas density $\rho^{\mathrm{G}}(t) = 3\, p^{\mathrm{G}}(t)$ and the velocity $\mathbf{v} = \mathbf{v}(\mathbf{x}, t)$ of the interface cell $\mathbf{x}$.
Inserting Eq. (8.8) in Eq. (8.7) gives:

$$\begin{aligned}
F_\alpha / A &= -n_\beta \sum_{i, \mathbf{n}\cdot\mathbf{e}_i < 0} f_i^{\mathrm{out}}(\mathbf{x}, t)\, (e_{i,\alpha} - v_\alpha)(e_{i,\beta} - v_\beta) \\
&\quad -n_\beta \sum_{i, \mathbf{n}\cdot\mathbf{e}_i \geq 0} \left[f_i^{\mathrm{eq}}(\rho^{\mathrm{G}}, \mathbf{v}) + f_{\bar{\imath}}^{\mathrm{eq}}(\rho^{\mathrm{G}}, \mathbf{v}) \ - f_{\bar{\imath}}^{\mathrm{out}}(\mathbf{x}, t) \right] (e_{i,\alpha} - v_\alpha)(e_{i,\beta} - v_\beta) \\
&= -n_\beta \sum_{i, \mathbf{n}\cdot\mathbf{e}_i \geq 0} \left[f_i^{\mathrm{eq}}(\rho^{\mathrm{G}}, \mathbf{v}) + f_{\bar{\imath}}^{\mathrm{eq}}(\rho^{\mathrm{G}}, \mathbf{v}) \right] (e_{i,\alpha} - v_\alpha)(e_{i,\beta} - v_\beta) \\
&= -n_\beta \sum_i f_i^{\mathrm{eq}}(\rho^{\mathrm{G}}, \mathbf{v})(e_{i,\alpha} - v_\alpha)(e_{i,\beta} - v_\beta) \\
&\overset{[219]}{=} -n_\beta\, \delta_{\alpha\beta}\, p^{\mathrm{G}} \\
&= -n_\alpha \cdot p^{\mathrm{G}} \qquad (8.9)
\end{aligned}$$

The resulting fluid pressure F_α/A has the same value as the gas pressure p^{G} but with the opposite direction $-\mathbf{n}$. It is important to note that not only the missing distribution functions are reconstructed but all distribution functions with $\mathbf{e}_i \cdot \mathbf{n} \geq 0$ (see Fig. 8.3, right). At first glance, part of the information seems to be disregarded since some of the distribution functions coming from neighboring interface cells are not taken into account. In order to resolve this apparent contradiction it is instructive to remind that this seeming discrepancy originally results from the treatment of the interface cells. They are more or less treated as fluid cells with a full set of distribution functions defining the fluid. This brings some kind of asymmetry into the interface treatment. The above procedure restores in some sense symmetry since interface cells are sometimes treated as gas or as fluid depending on the actual interface orientation.

8.2.4 Collision and Gravity

After completion of the whole set of distribution functions, the new density ρ and velocity $\mathbf{v}$ can be calculated. The outgoing distribution functions are given by (see Sect. 7.3.2.4)

$$f_i^{\mathrm{out}}(\mathbf{x},t) = f_i^{\mathrm{in}}(\mathbf{x},t) - \frac{1}{\tau}\left(f_i^{\mathrm{in}}(\mathbf{x},t) - f_i^{\mathrm{eq}}(\rho,\mathbf{v})\right) + \epsilon(\mathbf{x})\, 3\, w_i\, \rho\, [(\mathbf{e}_i - \mathbf{v}) + 3(\mathbf{e}_i \cdot \mathbf{v})\, \mathbf{e}_i] \cdot \mathbf{g} \qquad i = 1, \ldots, b, \quad \mathbf{x} \in \mathrm{I}. \tag{8.10}$$

Similar to the volume of fluid method gravity is weighted in Eq. (8.10) by the fluid volume fraction ϵ of the cell. This is the same approach as in [86, 87].

The reconstruction of the missing distribution functions at interface cells is solely based on the balance of distribution functions. Gradient information is not explicitly used. As a result, the interface reconstruction procedure is not time consuming which is especially important for the underlying foam formation problem. In order to prove the validity of our algorithm, 2D (bubble advection, capillary waves) and 3D (bubble advection and rising bubble) numerical experiments have been performed where analytical solutions are available, for details see [154, 155, 162].

9 Lattice Boltzmann Model for Foam Evolution

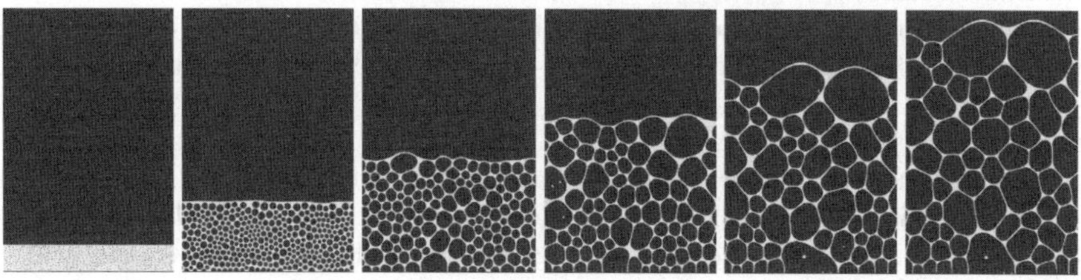

The availability of a LB algorithm for free surface flow is an important element to model foam evolution but not sufficient. What is still missing is the description of the basic components of foam, the gas bubbles. In the following, the ingredients which transform the LBM for free surfaces to a method to model foam evolution processes are described. This model extension comprises gas release and diffusion, the local curvature, the foam stabilizing disjoining pressure and – last but not least – bubble coalescence. The resulting LBM for the simulation of foam evolution processes shows a variety of complex phenomena such as rapid structure transformations, topological changes (T1- or T2-processes), avalanches, bubble sorting, drainage, decay, etc., which simply follow without further approximation from hydrodynamics coupled with surface energy effects and bubble coalescence.

9.1 Gas Bubbles and Gas Transport

9.1.1 Gas Bubbles

Bubbles are defined by interconnected gas cells and their neighboring interface cells. All cells belonging to bubble B_j share the identical information: the bubble volume V_j and the gas content n_j. The bubble pressure, p_j, follows from the ideal gas equation, Eq. (4.3). The amount of gas in a bubble is changed by diffusion processes. Bubble volume changes follow from fluid dynamics driven by pressure gradients and viscous forces.

The bubble volume V_j is given by the sum of the gas fractions of all cells belonging to bubble j,

$$V_j = \sum_{\mathbf{x}\in B_j \ \wedge \ \mathbf{x}\in I} (1 - \epsilon(\mathbf{x})) \ \Delta V + \sum_{\mathbf{x}\in B_j \ \wedge \ \mathbf{x}\in G} \Delta V, \tag{9.1}$$

where ΔV denotes the cell volume (generally, $\Delta V = 1$).

9.1.2 D2Q4 Model for Gas Transport [238]

Foam evolution depends on the amount of gas released by the blowing agent during a certain time period. The blowing agent particles are described with the help of a volume source, $Q = Q(\mathbf{x}, \mathbf{t})$. The gas is dissolved in the melt and diffuses in the direction of concentration sinks, the bubbles, and inflates them. Altogether, a convection–diffusion equation with space-time dependent volume source has to be solved.

The convection–diffusion equation is a linear, scalar differential equation, see Eq. (4.7), and thus much easier to handle than the non-linear, vectorial NSE. In order to guarantee isotropy, a LBM with only four velocities is already sufficient. A two-dimensional LBM (D2Q4 model, see Fig. 7.9 and Table 7.1) on a square grid is considered. The gas concentration c is given by the sum of the set of distribution functions $g_i(\mathbf{x}, t)$,

$$c(\mathbf{x}, t) = \sum_{i=1}^{4} g_i(\mathbf{x}, t). \tag{9.2}$$

The evolution equation reads:

$$g_i(\mathbf{x} + \mathbf{e}_i, t + 1) = g_i(\mathbf{x}, t) - \frac{1}{\tau_\mathrm{D}} \left[g_i(\mathbf{x}, t) - g_i^{\mathrm{eq}}(\mathbf{x}, t)\right] + \frac{1}{4} Q(\mathbf{x}, t) \qquad i = 1, ..., 4 \tag{9.3}$$

where τ_D and Q denote the relaxation time for diffusion and the source term, respectively. Coupling between the convection–diffusion equation and the NSE is via the velocity $\mathbf{v}$. The equilibrium functions g_i^{eq} are given by[1]

$$g_i^{\mathrm{eq}}(\mathbf{x}, t) = g_i^{\mathrm{eq}}(c(\mathbf{x}, t), \mathbf{v}(\mathbf{x}, t)) = \frac{c}{4} \left(1 + \frac{\mathbf{e}_i \cdot \mathbf{v}}{c_\mathrm{D}^2}\right) \tag{9.4}$$

with

$$c\,\mathbf{v} = \sum_{i=1}^{4} g_i\, \mathbf{e}_i. \tag{9.5}$$

The velocity c_D is given by $c_\mathrm{D}^2 = \frac{1}{2}$ and the diffusion constant D is related to the relaxation time through

$$D = c_\mathrm{D}^2 \left(\tau_\mathrm{D} - \frac{1}{2}\right) = \frac{1}{2} \left(\tau_\mathrm{D} - \frac{1}{2}\right). \tag{9.6}$$

[1]The linearity allows to simplify the equilibrium functions by omitting the quadratic terms.

9.1.3 Boundary Conditions

The interface with respect to the diffusion problem is different from that for the fluid dynamic problem due to the different neighborhoods, v. Neumann and Moore for the D2Q4 and D2Q9 approach, respectively (see Fig. 7.6). Analogous to the fluid flow problem, distribution functions g_i originating from the gas are not defined. They have to be reconstructed on condition that the appropriate Dirichlet boundary condition given by Sieverts' law, Eq. (4.9),

$$c|_{\Gamma} = S\sqrt{p_j} = c_{\mathrm{B_j}}, \tag{9.7}$$

is fulfilled. The gas concentration which is in equilibrium with bubble j is denoted by $c_{\mathrm{B_j}}$.

Due to the discrete lattice we use the much simpler condition,

$$c(\mathbf{x},t)|_{\mathbf{x}\in\Gamma_j} = S\sqrt{p_j} = c_{\mathrm{B_j}}, \tag{9.8}$$

where the set of interface cells belonging to bubble j is denoted by Γ_j, see Fig. 9.1.[2]

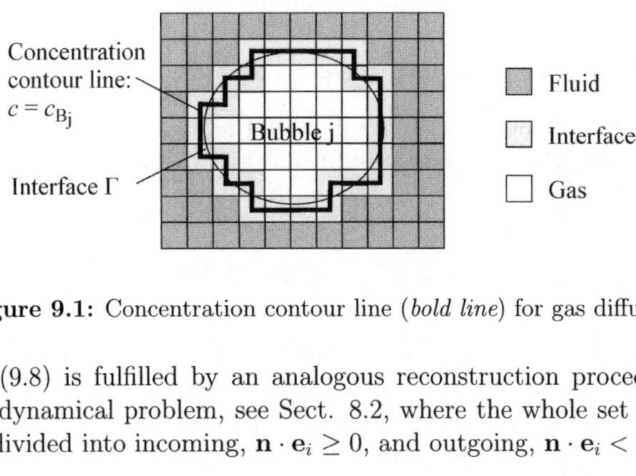

Figure 9.1: Concentration contour line (*bold line*) for gas diffusion.

Equation (9.8) is fulfilled by an analogous reconstruction procedure employed for the fluid dynamical problem, see Sect. 8.2, where the whole set of distribution functions is divided into incoming, $\mathbf{n}\cdot\mathbf{e}_i \geq 0$, and outgoing, $\mathbf{n}\cdot\mathbf{e}_i < 0$, ones:

$$g_i^{\mathrm{out}}(\mathbf{x}-\mathbf{e}_i,t) = g_i^{\mathrm{eq}}(c_{\mathrm{B_j}},\mathbf{v}) + g_{\bar{i}}^{\mathrm{eq}}(c_{\mathrm{B_j}},\mathbf{v}) - g_{\bar{i}}^{\mathrm{out}}(\mathbf{x}+\mathbf{e}_i,t) \qquad \forall i\colon\ \mathbf{n}\cdot\mathbf{e}_i \geq 0. \tag{9.9}$$

The last relation is equivalent with

$$g_i^{\mathrm{in}}(\mathbf{x},t+1) = g_i^{\mathrm{eq}}(c_{\mathrm{B_j}},\mathbf{v}) + g_{\bar{i}}^{\mathrm{eq}}(c_{\mathrm{B_j}},\mathbf{v}) - g_{\bar{i}}^{\mathrm{in}}(\mathbf{x},t+1) \qquad \forall i\colon\ \mathbf{n}\cdot\mathbf{e}_i \geq 0. \tag{9.10}$$

[2]This procedure appears quite crude but turns out to be acceptable in view of the much larger uncertainty due to blowing agent decomposition, see Appendix A.2.

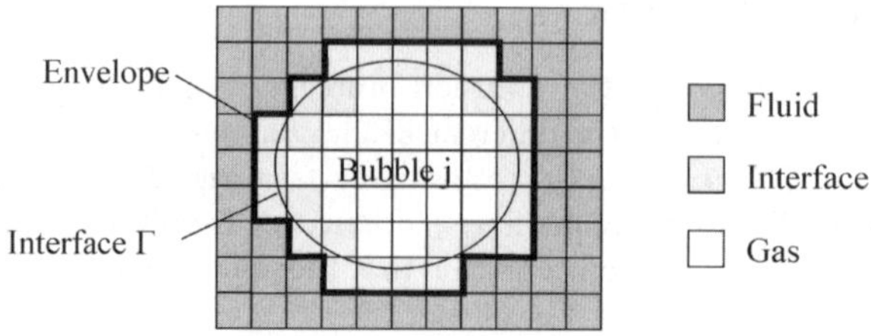

Figure 9.2: Envelope for the gas exchange of the bubble with the fluid metal.

It is important to note, that the incoming inverse distribution function are used for reconstruction. This is different from the reconstruction procedure for the fluid dynamical case where the inverse outgoing ones are exploited.

The concentration at interface cells is thus given by

$$c(\mathbf{x}, t+1)|_{\mathbf{x}\in\Gamma_j} = \sum_{i=1}^{4} g_i^{\text{in}}(\mathbf{x}, t+1) = \sum_{i=1}^{4} g_i^{\text{eq}}(c_{\text{B}_\text{j}}, \mathbf{v}) = c_{\text{B}_\text{j}}, \tag{9.11}$$

which is exactly the imposed boundary condition Eq. (9.8).

9.1.4 Gas Current into Bubbles

The amount of gas streaming into a bubble during one time step is a function of the interface length, the diffusion constant and the concentration gradients and is estimated by

$$\Delta n_j = \sum_{\mathbf{x}\in\Gamma_j} \left[-c(\mathbf{x}, t)\Delta\epsilon + \sum_{\mathbf{x}+\mathbf{e}_i\in F} [g_{\bar{i}}(\mathbf{x}+\mathbf{e}_i, t) - g_i(\mathbf{x}, t)] \right]. \tag{9.12}$$

The first term accounts for the pure convective part of gas transport which does not contribute to the gas gain in the bubble. It has to be subtracted from the second term which is the total gas current across the fluid–gas interface. This interface is approximated by the outer envelope of the bubble, see Fig. 9.2. For very small bubbles, the gas current is overestimated by Eq. (9.12).

9.1.5 Bubble Coalescence

Cell wall rupture events lead to bubble coalescence. Technically, cell wall rupture is initiated if an interface cell is claimed by two distinct bubbles. In this case, bubble coalescence is the consequence. Coalescence is accomplished by transforming the relevant interface cell into a gas cell and by merging the bubble properties. In detail, the bubble volumes and the amounts of substance are summed up. Subsequently, the pressure is recalculated and the new united bubble information is subscribed to all cells belonging to the former two bubbles. By applying this rupture criterion, the determination of the rupture events is not based on an exactly prescribed cell wall rupture thickness. Rupture events can take place at any thickness between 0 and the lattice spacing Δx.

9.2 Interfacial Forces

The variety of phenomena observed during foam evolution such as drainage, avalanches, topological changes, etc., are the outcome of the interplay between hydrodynamics and interfacial forces. There are two kinds of interfacial forces which have to be taken into account for the underlying foam formation problem, the capillary force and the disjoining pressure. The capillary force has the tendency to destroy the foam structure whereas the disjoining pressure shows a stabilizing effect.[3]

9.2.1 Capillary Force

The capillary force, see Eq. (4.5), results from the energy necessary to create an interface. Their effect is always to minimize the surface area. The higher the mean curvature κ is, the higher is the resulting capillary force. The effect of the capillary force is treated as a local modification of the gas pressure p^{G}:

$$\text{2D:}\quad p^{\mathrm{G}} \longmapsto p^{\mathrm{G}} - \sigma\,\kappa \qquad\qquad \text{3D:}\quad p^{\mathrm{G}} \longmapsto p^{\mathrm{G}} - 2\,\sigma\,\kappa, \tag{9.13}$$

where κ and σ denote the curvature and the surface tension, respectively.

The curvature κ of a curve $\mathbf{r}(s)$ is defined as the change of the direction with unit length [96],

$$\kappa = \left\| \frac{d^2\mathbf{r}}{ds^2} \right\| \qquad \text{with} \qquad ds = \|d\mathbf{r}\|. \tag{9.14}$$

For a plane curve the curvature is identical with the divergence of the normal vector,

$$\kappa = |\nabla \cdot \mathbf{n}|\,, \tag{9.15}$$

where $\mathbf{n}$ is the normal vector with $\|\mathbf{n}\| = 1$.

[3]The introduction of interfacial forces destroys one very important feature of the underlying cellular automaton, its locality. The direct neighbors of an interface cell are not sufficient to calculate the relevant interfacial forces and a much more extended neighborhood is necessary. Thus, parallelization of the computer code, which is often said to be a very important property of the LBM, is not straightforward [154, 155].

The curvature of a surface is characterized by its mean curvature κ,

$$\kappa = \frac{1}{2}\left(\kappa_1 + \kappa_2\right) \tag{9.16}$$

where κ_1 and κ_2 denote the two principal curvatures.

The determination of the curvature is a difficult and time consuming task! The difficulty is founded on the second derivative which has to be estimated. Since the curvature has to be calculated for each time step, the underlying algorithm should not be too time consuming. On the other hand, the curvature has to be determined as exact as possible since the capillary forces are the basis for the complex foam behavior.

There are different strategies to calculate κ from the information available on the LB lattice. None of these strategies has proved to be perfect but all are time consuming. Attempts to calculate the curvature only from the nearest neighbors give unusable results. For acceptable results, a larger neighborhood has to be considered. In two dimensions, a template sphere model has shown to suit best, for details see Appendix B.3.[4]

9.2.2 Disjoining Pressure

Foam stabilization is accomplished by a disjoining pressure which is a function of the local wall thickness d, see Eq. (7.1). The disjoining pressure Π is analogously treated as the capillary forces in Eq. (9.13):

$$\text{2D:}\quad p^{\mathrm{G}} \longmapsto p^{\mathrm{G}} - \sigma \cdot \kappa - \Pi \qquad\qquad \text{3D:}\quad p^{\mathrm{G}} \longmapsto p^{\mathrm{G}} - 2\,\sigma \cdot \kappa - \Pi. \tag{9.17}$$

For parallel interfaces, the meaning of d is clear whereas it has to be defined for non-parallel ones. The local wall thickness belonging to interface cell i is assumed to be the diameter of the largest sphere which just fits in between the two interfaces and which touches the interface at interface cell i, see Fig. 9.3(a).

In a first step, the diameter of the largest sphere is estimated by the distance from one interface to the neighboring interface in negative normal direction.[5] The calculation of this distance is performed in two steps:

1. Determination of the nearest interface cell in negative normal direction which belongs to bubble j with $j \neq k$, Fig. 9.3(b).[6]

[4]In three dimensions, we use a marging cube algorithm for surface triangulation and calculate the curvature with the help of normal vectors [154, 155].

[5]Large deviations only occur if the normals of two neighboring interfaces are strongly different.

[6]Effective methods to determine the cells lined up along the normal vector $\mathbf{n}$ developed in the context of ray tracing problems, see e. g. Cohen [49].

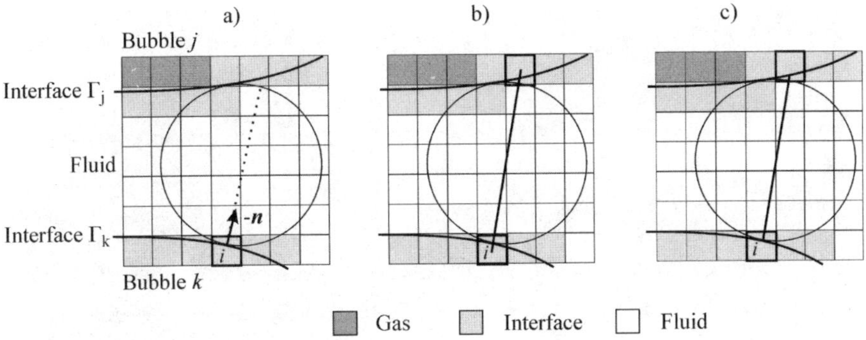

Figure 9.3: Determination of the local wall thickness d. (**a**) Definition. (**b**) Determination of the nearest interface cell by propagation along the negative normal direction. (**c**) Estimate of the distance by taking into account the fluid fraction content ϵ.

2. Determination of the distance of the two interface cells and correction of this distance by taking into account the respective fluid volume contents of the interface cells, Fig. 9.3(c).

The first step has to be numerically efficient. The second step, where the actual fluid volume content is taken into account, is very important. Without correction the distance might jump up to one lattice spacing. The correction makes d and Π continuous and thereby avoids artificial fluctuations [223]. Generally, the range of the disjoining pressure, d_π, is assumed to be three lattice units.

9.3 Foam Formation Phenomena

Which complex phenomena are covered by the numerical approach? The physical model takes into account surface tension, viscosity, gravity, gas diffusion and a phenomenological disjoining pressure. In addition, cell wall rupture is, even if simplified, taken into account. Thus, most of the phenomena observed during foam evolution are the outcome of the model without any further approximations or model assumptions.

9.3.1 Foam Expansion and Decay

The complete foam evolution process starting from nucleation, bubble growth, aging, and eventually decay, is described (Fig. 9.4).

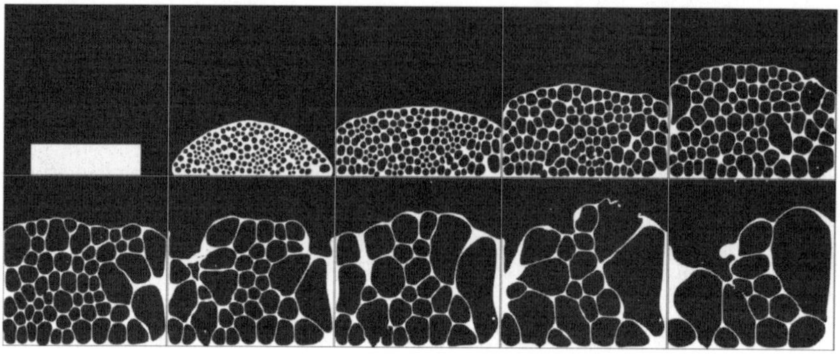

Figure 9.4: Foam expansion and decay. (2D-LBM simulation)

The foaming result is governed by the nuclei density, the surface tension, the viscosity, the disjoining pressure, the gas release rate and, last but not least, by the confining mold geometry, see Fig. 9.5. Complex mold geometries represent neither a further difficulty nor demand a much higher computing time in the framework of the LB approach.

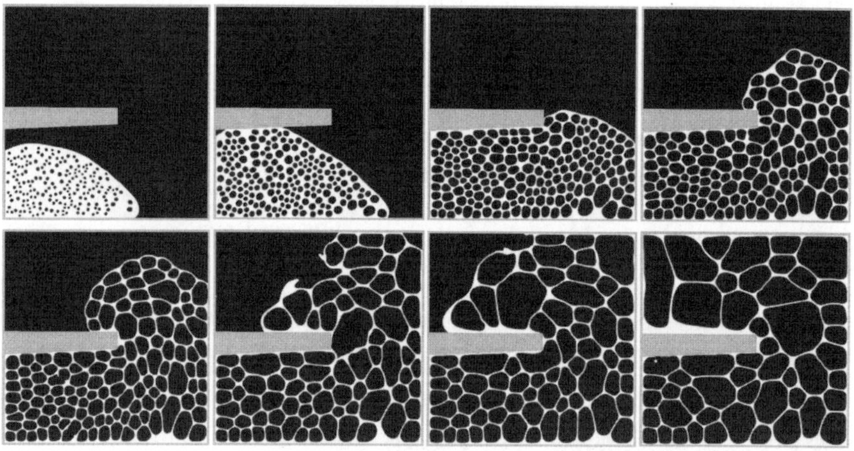

Figure 9.5: Foam formation in a complex mold. (2D-LBM simulation)

9.3.2 Topological Rearrangement Processes

Topological bubble rearrangement processes occur due to several reasons such as spacial restrictions owing to growth, cell wall coalescence events or external mechanical forces. Figure 9.6 shows a T1-process during foam shearing. It is important to pro-

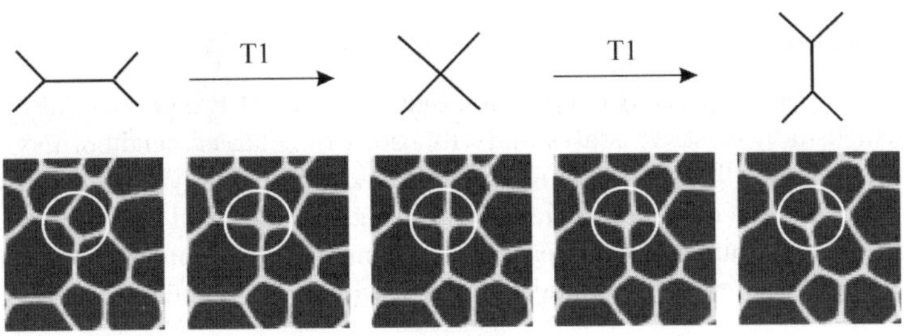

Figure 9.6: T1-process (*white circle*). *Top*: Schematic process. *Bottom*: T1-process during shearing of a foam structure. (2D-LBM simulation)

nounce that topological rearrangement processes are the outcome of the underlying model and not the result of additional model assumptions.

The numerical model has also the potential to model T2-processes where small bubbles vanish as a consequence of gas loss due to the higher equilibrium gas pressure in small bubbles compared to large ones. This phenomenon is sometimes called Ostwald ripening. For metal foam this process is quite slow due to the thick cell walls of about 10^2µm [15, 146].[7]

9.3.3 Cell Wall Rupture

Cell wall rupture initiates a succession of processes, see Fig. 9.7. The cell wall is pulled back to the Plateau borders driven by capillary forces. Further material redistribution processes are required to adapt the cell walls to the new cell configuration. The time scale for these redistribution processes is proportional to the capillary velocity $v_{ca} = \sigma/\eta$, see Sect. 5.2.1. Material redistribution processes are no longer possible if v_{ca} falls behind the expansion velocity of the foam.

[7] The situation is different for aqueous foams where the wall thickness is several orders of magnitude smaller. The evolution of standing aqueous foams is very much influenced by diffusion induced coarsening [72, 186, 199].

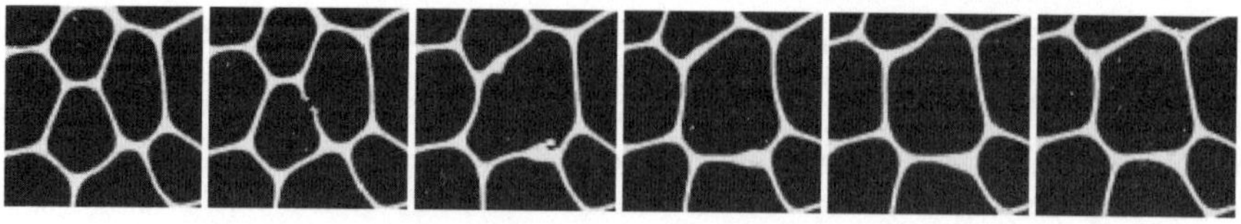

Figure 9.7: Cell wall rupture event. (2D-LBM simulation)

9.3.4 Avalanches

Generally, cell wall rupture is not a local event but releases a high amount of elastic energy which induces stress states and vibrations of a larger neighborhood of the original rupturing cell wall. These mechanical loads are possibly able to induce new rupture events in the neighborhood. The resulting cascades lead to rapid coarsening of the cellular structure where larger volumes completely reorganize. Located at the foam surface, cascades lead to rapid gas loss initiating foam collapse, see Fig. 9.8.

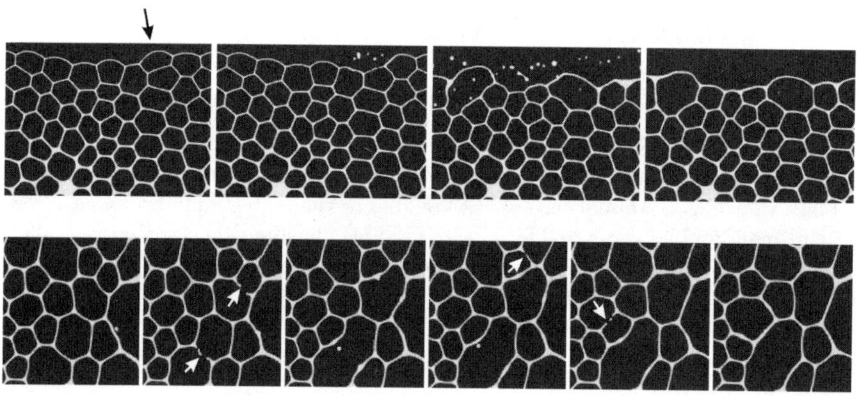

Figure 9.8: Avalanches. *Top*: Bubble coalescence cascade at the foam surface which leads to gas loss and initiates decay. *Bottom*: Foam coarsening by a succession of internal bubble coalescence events. (2D-LBM simulation)

9.3.5 Drainage

Drainage denotes the rearrangement of fluid under the influence of capillary forces and gravity. Capillary forces push fluid from cell walls into Plateau borders whereas gravity makes the fluid running out of the foam structure. The equilibrium state is characterized by the balance between gravity, capillary forces and the disjoining pressure. As a result, a gradient of the curvature radius R_{Pl} of the Plateau borders develops, i. e. R_{Pl} decreases with increasing height, see Fig. 9.9.

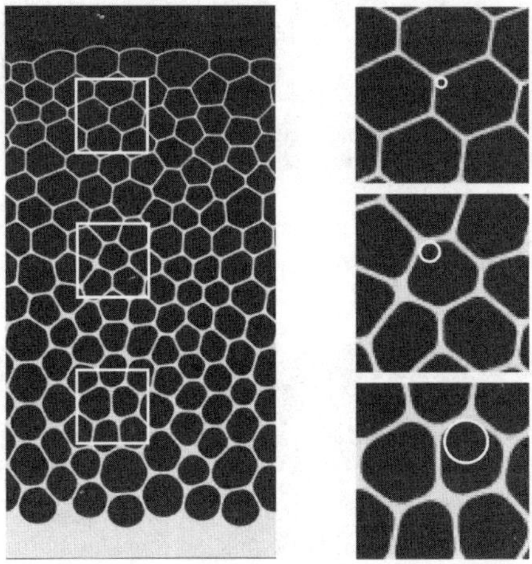

Figure 9.9: Foam structure under gravity. *Left*: Overall view with drainage zone at the bottom. *Right*: Magnifications at different heights and inscribed circles with radius R_{Pl}. (2D-LBM simulation)

9.3.6 Pulsed Drainage

Adding a certain amount of fluid on top of a foam leads to a phenomenon called pulsed drainage. The pulse spreads out while moving downwards through the foam. For the example depicted on Fig. 9.10, the fluid pulse causes a phase transition from a dry foam to a fluid foam and vice versa. As a result, the cellular structure is completely rearranged.

9.3.7 Experimental vs Numerical Structures

Figure 9.11 shows different aluminum foam structures compared to 2D numerical results. The differences of the structures occur due to differences of the viscosity, expansion velocity, relative density or the disjoining pressure. Although two-dimensional, the simulated cell structures show the main characteristics of real foams.

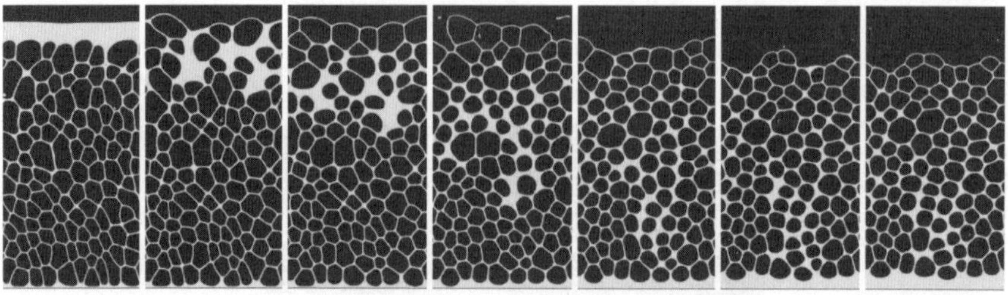

Figure 9.10: Pulsed drainage. The fluid pulse spreads out and causes the cellular structure to reorganize. (2D-LBM simulation)

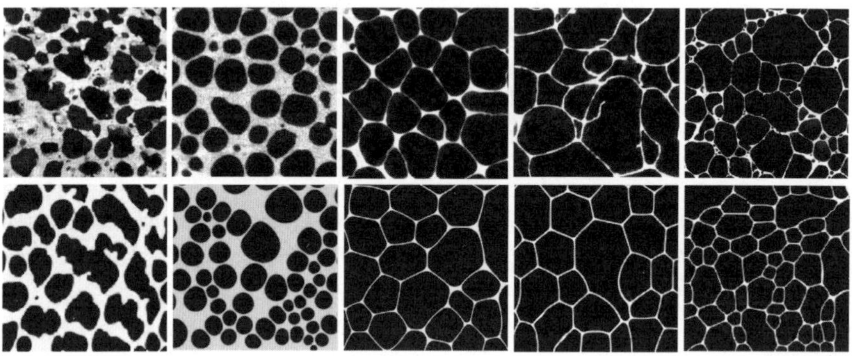

Figure 9.11: Experimental aluminum foam structures (*top*) vs 2D numerical structures (*bottom*) [160]. (2D-LBM simulation)

9.3.8 3D Foam Simulation

3D simulations are much more time consuming than 2D ones. Analogous to 2D, drainage, topological rearrangements, cascades, etc., are observed without further model assumptions. Some examples are depicted in Fig. 9.12.

Other phenomena such as foam rheology [223] or bubble sorting phenomena are also described by the numerical approach.

9.3.9 Limitations

The numerical model shows limitations which might be essential for certain classes of foam.

Problems emerge for foams with very thin cell walls e. g. soap foam. In this case, the limitations result from the enormous number of cells which would be necessary

Figure 9.12: 3D foams. *Top*: Free expansion. *Middle* and *bottom*: Expansion with periodic boundary conditions in x- and y-direction [155, 162].

to represent the cell walls. Details of the cell wall rupturing process are also not captured. A better description would require a much higher numerical resolution. However, we do not expect strong general impact on the evolution of the cellular structure. In addition, the effect of surface active substances which locally change the surface tension is – up to now – not taken into account.

A Gas Supply and Gas Diffusion

A.1 Generalized Johnson-Mehl-Avrami Approach

Aim of this section is to describe a theoretical approach to apply JMA theory (see Sect. 2.5.2.2), which is generally used at constant temperature, to experiments with constant heating rate. The following description is based on the PhD thesis by Kempen [135].

A.1.1 Theory

The JMA approach, Eq. (2.5), has to be generalized in order to apply it to the experimental situation of a constant heating rate $\dot{T}$ with

$$T(t) = T_0 + \dot{T}\,t, \tag{A.1}$$

where T_0 denotes the initial temperature.

It is assumed that the thermal history, i. e. the path in the time-temperature-diagram, determines the degree of transformation (of the blowing agent MgH_2). A path variable β, which is a function of the thermal history describes the degree of transformation [135, 180]:

$$f(t) = \exp\left(-\beta^n\right). \tag{A.2}$$

The dependence of the path variable β on the thermal history can be described as a time integral of a constant $k(T(t))$, where the time dependence is only implicit through the temperature:

$$\beta(t) := k_0 \int_0^t \exp\left(-\frac{E_\mathrm{A}}{RT(\tilde{t})}\right)\mathrm{d}\tilde{t} \quad \overset{\text{Eq. (A.1)}}{=} \quad k_0 \int_0^t \exp\left(-\frac{E_\mathrm{A}}{R(T_0+\dot{T}\tilde{t})}\right)\mathrm{d}\tilde{t}. \tag{A.3}$$

Equation Eq. (A.3) is based on additivity, which is only valid if the transformation mechanism does not change throughout the temperature range of interest. We follow the approach in [135] and substitute the time variable t by

$$\hat{t} = t + T_0/\dot{T} \; = \; T/\dot{T}. \tag{A.4}$$

The time variable $\hat{t}$ is the time which would be necessary to heat the sample from $T = 0$ K to the temperature T with constant heating rate $\dot{T}$. With this new time variable, Eq. (A.3) reads [135]:

$$\beta(\hat{t}) = k_0 \int_{T_0/\dot{T}}^{\hat{t}} \exp\Big(-\frac{E_A}{R\dot{T}\tilde{t}}\Big)\mathrm{d}\tilde{t} \approx k_0 \int_0^{\hat{t}} \exp\Big(-\frac{E_A}{R\dot{T}\tilde{t}}\Big)\mathrm{d}\tilde{t} \approx \frac{k_0\,\dot{T}\,R\hat{t}^2}{E_A}\exp\Big(-\frac{E_A}{R\dot{T}\hat{t}}\Big). \quad (A.5)$$

Back substitution of $\hat{t}$ by $T/\dot{T}$ results in

$$\beta(T) = \frac{k_0\,RT^2}{E_A\,\dot{T}} \exp\Big(-\frac{E_A}{RT}\Big). \quad (A.6)$$

A.1.2 Material Constants for MgH_2

There are three constants which have to be experimentally determined, E_A, n and k_0, compare Eq. (2.5) and Eq. (k-for-JMA). The apparent activation energy was found by Andreasen et al. [3] with the help of isothermal experiments: $E_A = 160$ kJ/mol. The Avrami exponent n and the constant k_0 are determined by fitting the theoretical decomposition data, Eq. (A.6), to own decomposition experiments[1], see Fig. A.1.

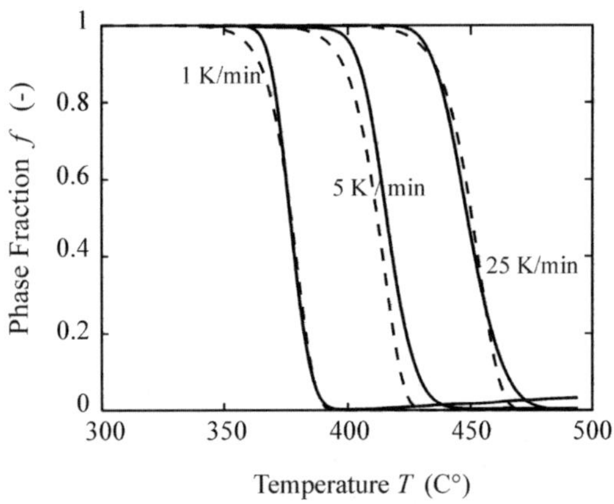

Figure A.1: Decomposition of MgH_2 at heating rates of 1, 5 and 25 K/min. *Full lines*: Experimental data (thermo-gravimetry). *Dashed lines*: Theoretical data fitted by applying Eq. (A.2), Eq. (A.6) and Table A.1.

[1]The decomposition of MgH_2 was measured by thermo-gravimetry: NETZSCH STA 409C, temperature range: 20–500 °C, argon gas.

The temperature at the onset of decomposition is determined by k_0 whereas the steepness of the decline is given by n.[2] The JMA theory is in fact able to describe the decomposition and its dependence on the heating rate. The physical parameters are listed in Table A.1.

Table A.1: Parameters for the decomposition of MgH_2 on basis of JMA theory.

E_A	n	k_0
160 kJ/mol	3	$4.63 \cdot 10^9\,s^{-1}$

A.2 Gas Diffusion

There are several situations where the exact description of gas diffusion is essential for foam evolution:

- Very low density of blowing agent particles.
- Ostwald ripening. The different capillary pressures of bubbles with different diameter lead to gas currents from smaller bubbles to larger ones. Although very important for aqueous foams, Ostwald ripening has been demonstrated to be irrelevant for metal foams [15, 146].

During IFM, gas is provided by quite large (10–100 µm) MgH_2 particles whose role is to act as bubble nucleation sites and to release gas directly into bubbles, see Fig. A.2, left. The role of gas diffusion appears to be quite moderate due to the direct contact with the bubbles and the small diffusion length calculated from the diffusion constant. The most exact way to describe this situation would be to consider each blowing agent particle as individual blowing agent source. Such a procedure would strongly exceed the numerical capacities.

In order to mimic the real situation the individual blowing agent particles are considered as homogeneous volume source, see Fig. A.2, right. In addition, the diffusion constant is strongly increased to make the gas diffusing very quickly to the bubble nuclei. This procedure has a similar effect as individual blowing agent particles releasing their gas directly into bubbles. During foam expansion, the number of blowing agent particles present at bubble interfaces increases due to coalescence events. Thus, differences between the treatment of individual blowing agent particles and the homogeneous volume source more and more diminish with increasing

[2] The Avrami exponent obtained in literature varies between $n = 2$ and $n = 4$ [69, 70].

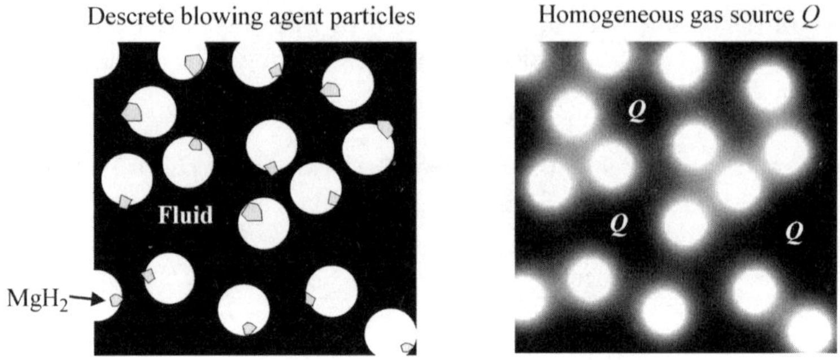

Figure A.2: Gas release and diffusion (schematic). *Left*: Real process. Blowing agent particles acting as bubble nuclei and releasing their gas directly into bubbles. *Right*: Modeling. Homogeneous volume source and high diffusion constant.

porosity. In order to imitate the real situation, the physical material parameters[3] have to be adapted. We demand:

1. $Q^\star \cdot n_{\mathrm{steps}} \gg S^\star$ where n_{steps} denotes the number of time steps,
2. $Q^\star \ll S^\star$ and
3. $\sqrt{4\,D^\star \cdot n_{\mathrm{steps}}} \gg 1$.

Condition 1 guarantees that the total amount of released gas is much larger than the amount of gas which is dissolved in the fluid. Condition 2 and 3 ensure that the gas does not accumulate in the fluid but diffuses rapidly to the bubbles.

[3]Dimensionless parameters are marked with $\star$, see also Appendix C.

B Miscellany

B.1 Foaming Pressure

The aim of this section is to estimate the foaming pressure during IFM of aluminum. In a first step, the amount and volume of hydrogen gas evolving during IFM is calculated. The amount of hydrogen gas n_{H_2} contained in 1 g MgH_2 follows from

$$n_{H_2} = \frac{m_{MgH_2}}{M_{MgH_2}} = \frac{1\,\mathrm{g}}{26.32\,\mathrm{g/mol}} = 0.038\,\mathrm{mol}, \tag{B.1}$$

where m_{MgH_2} and M_{MgH_2} denote the mass and the molar mass of MgH_2, respectively.

The volume of hydrogen gas V_{H_2} contained in 1 g MgH_2 is given by the ideal gas law,

$$V_{H_2} = n_{H_2} \cdot \frac{RT}{p} = \frac{2750\,\mathrm{cm}^3\,\mathrm{bar}}{p}, \tag{B.2}$$

where $T = 600°\mathrm{C}$ is assumed.

Only part of the total blowing agent decomposes due to the chilling effect of the walls. In a simple model, the casting is divided into a compact skin where no decomposition of the blowing agent takes place and a cellular core where complete decomposition is assumed. The total mass of decomposed MgH_2 is given by the mass of the cellular core m_{core} and the mass fraction ϵ_{m} of the blowing agent,

$$m_{MgH_2} = \epsilon_{\mathrm{m}} \cdot m_{\mathrm{core}} = \epsilon_{\mathrm{m}} \cdot \rho_{\mathrm{rel,core}} \cdot \rho_{\mathrm{Al}} \cdot V_{\mathrm{core}} \tag{B.3}$$

with V_{core}: volume of the cellular core, ρ_{Al}: density of aluminum and the relative density of the core,

$$\rho_{\mathrm{rel,core}} = 1 - (1 - \rho_{\mathrm{rel}}) \cdot \frac{d}{d_{\mathrm{core}}}, \tag{B.4}$$

where d and d_{core} denote the thickness of the casting and the core, respectively.

The gas volume V_{H_2} resulting from the blowing agent is

$$V_{H_2} = m_{MgH_2} \cdot \frac{2750}{p} \frac{\mathrm{cm}^3\,\mathrm{bar}}{\mathrm{g}} = \epsilon_{\mathrm{m}} \cdot \rho_{\mathrm{rel,core}} \cdot \rho_{\mathrm{Al}} \cdot V_{\mathrm{core}} \cdot \frac{2750}{p} \frac{\mathrm{cm}^3\,\mathrm{bar}}{\mathrm{g}}. \tag{B.5}$$

On the other hand, V_{H_2} is determined by the relative density of the cellular core,

$$V_{H_2} = V_{\mathrm{core}} \cdot (1 - \rho_{\mathrm{rel,core}}). \tag{B.6}$$

The resulting **foaming pressure** p follows from Eqs. (B.5) and (B.6):

$$\begin{aligned} p &= \epsilon_\mathrm{m} \cdot \frac{\rho_\mathrm{rel,core}}{(1-\rho_\mathrm{rel,core})} \cdot \rho_\mathrm{Al} \cdot 2750\,\frac{\mathrm{cm^3\,bar}}{\mathrm{g}} \\ &= \epsilon_\mathrm{m} \cdot \frac{\left[1-(1-\rho_\mathrm{rel}) \cdot \frac{d}{d_\mathrm{core}}\right]}{(1-\rho_\mathrm{rel}) \cdot \frac{d}{d_\mathrm{core}}} \cdot \rho_\mathrm{Al} \cdot 2750\,\frac{\mathrm{cm^3\,bar}}{\mathrm{g}}. \end{aligned} \tag{B.7}$$

B.2 Mean Cell Diameter

In the following, the 2D and 3D mean cell diameter are referred to as D_{2D} and D_{3D}, respectively. In order to obtain local information about the cell diameter, the 3D cell diameter is estimated from 2D μCT data by determining the 2D mean cell diameter and weighting it with an appropriate factor to account for the 3D effect.

B.2.1 2D Mean Cell Diameter

D_{2D} is determined by exploring 2D μCT information. It is supposed that the cell structure may be described by a mean cell diameter. In addition, the cells are assumed to show a perfect round shape. With these assumptions and the 2D μCT information, which comprises the total area A_{tot}, the total material area A_{mat} contained in A_{tot} and the internal gas-material interface length L_{tot} within A_{tot}, it is possible to estimate the mean cell diameter D_{2D} as follows:

$$D_{2D} = 4\,\frac{A_{tot} - A_{mat}}{L_{tot} - 4\,\sqrt{A_{tot}}\,\frac{A_{mat}}{A_{tot}}}. \tag{B.8}$$

The second term in the denominator takes into account that the boundaries of the 2D area are also included in L_{tot} and have to be subtracted. The factor $\frac{A_{mat}}{A_{tot}} = \rho_{rel}$ accounts for the fact that the boundary length decreases with decreasing relative density.

B.2.2 3D Mean Cell Diameter

The mean cell diameter in 2D is a mean value, see Fig. B.1:

$$D_{2D} = \frac{2}{D_{3D}} \int_0^{D_{3D}/2} D(x)dx. \tag{B.9}$$

With

$$x = \frac{D_{3D}}{2} \cdot \cos\alpha \qquad \text{and} \qquad D = D_{3D} \cdot \sin\alpha \tag{B.10}$$

and

$$dx = -\frac{D_{3D}}{2} \cdot \sin\alpha\; d\alpha, \tag{B.11}$$

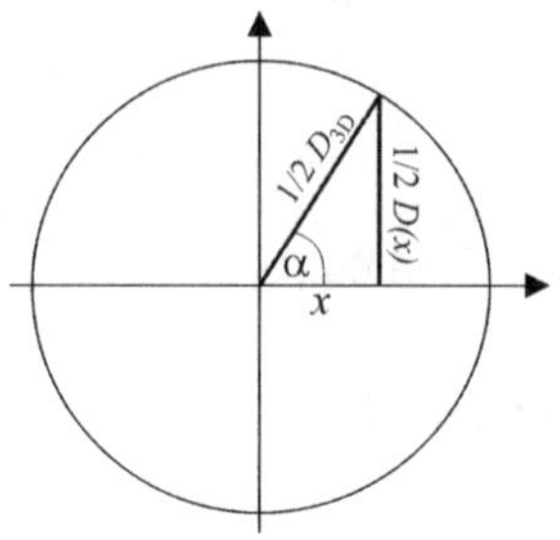

Figure B.1: Correlation between 2D and 3D mean cell diameter. The 2D diameter is a mean value due to different intersection planes.

the integration with respect to x is replaced by an integration over the angle α:

$$
\begin{aligned}
D_{2\mathrm{D}} &= -\frac{2}{D_{3\mathrm{D}}}\int_{\frac{\pi}{2}}^{0} \sin\alpha \cdot \sin\alpha \cdot \frac{D_{3\mathrm{D}}^2}{2}\, d\alpha = D_{3\mathrm{D}} \int_0^{\frac{\pi}{2}} \sin^2\alpha d\alpha \\
&= D_{3\mathrm{D}} \left[\frac{1}{2}\alpha - \frac{1}{4}\sin 2\alpha\right]_0^{\pi/2} = \frac{\pi}{4} \cdot D_{3\mathrm{D}} \qquad \text{(B.12)}
\end{aligned}
$$

B.3 Curvature Calculation in 2D: Template Sphere Model

The template sphere model is based on the observation that the curvature can be expressed in leading order by the portion of fluid volume enclosed by a template sphere centered on the interfacial points, see Fig. B.2 [223]:

$$\kappa = \frac{1}{R} = \frac{3\left(\frac{\pi r^2}{2} - A\right)}{r^3}. \tag{B.13}$$

where A is the area in the sphere covered with fluid.

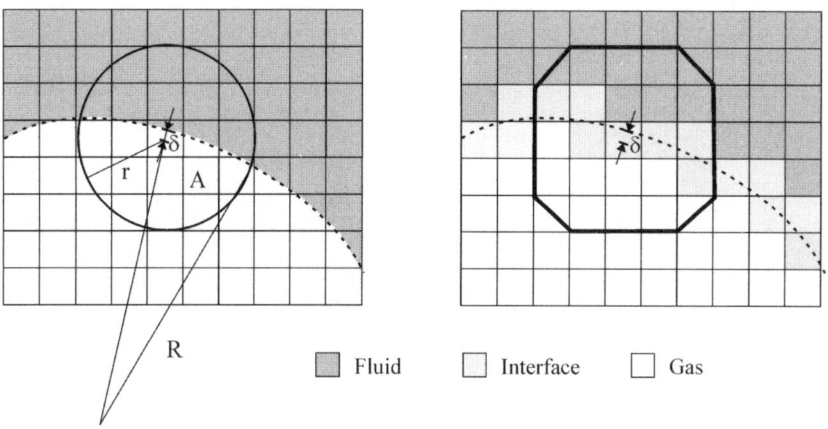

Figure B.2: Template sphere model. *Left*: Continuum situation. *Right*: Discrete lattice with 25-neighborhood. The distance between the center of the template sphere and the actual interface is referred to as δ.

On a discrete LB lattice, only neighborhoods which are similar to spheres, e. g. 25- or 36-neighborhood, can be defined. In order to reproduce the curvature as accurate as possible, Eq. (B.13) is modified in order to take into account that the interfacial point does generally not coincide with the center of the interface cell. With the distance between the center of the template sphere and the actual interface denoted with δ, the modification of Eq. (B.13) reads [223],

$$\kappa = \frac{1}{R} = \frac{3\left(\frac{\pi r^2}{2} - 2\,r\,\delta - A\right)}{r^3}. \tag{B.14}$$

A 25-neighborhood where the outermost cells are weighted by a factor $1/2$ has shown to represent the best trade-off between accuracy and computational work.

Generally, the mean curvature of a bubble is quite well reproduced. However, there might be strong fluctuations between neighboring interface cells especially in diagonal direction or before or after cell state transformation.

C Material Parameters

For the numerical simulation the material parameters, ν, σ, ρ, g, S, D, have to be expressed in dimensionless form. A time scale Δt, a length scale Δx, a mass scale Δm, and a scale for the amount of substance Δn have to be defined. The dimensionless quantities are marked with $\star$. The material parameters follow by multiplying the dimensionless quantities with the relevant scales:

$$\nu^\star = \nu \frac{\Delta t}{\Delta x^2} \quad \text{and} \quad \tau^\star = \frac{6\,\nu^\star + 1}{2} \tag{C.1}$$

$$\sigma^\star = \sigma \frac{\Delta t^2}{\Delta m} \tag{C.2}$$

$$\rho^\star = \rho \frac{\Delta x^3}{\Delta m} \tag{C.3}$$

$$g^\star = g \frac{\Delta t^2}{\Delta x} \tag{C.4}$$

$$S^\star = S \frac{\Delta x^3}{\Delta n} \sqrt{\frac{\Delta x\,\Delta t^2}{\Delta m}} \tag{C.5}$$

$$Q^\star = Q \frac{\Delta x^3}{\Delta n} \tag{C.6}$$

$$D^\star = D \frac{\Delta t}{\Delta x^2} \quad \text{and} \quad \tau_{\mathrm{D}}^\star = \frac{4\,D^\star + 1}{2} \tag{C.7}$$

In addition, pressure and density are correlated by

$$p = \frac{1}{3}\,\rho\,\frac{\Delta x^2}{\Delta t^2}. \tag{C.8}$$

For a given length scale Δx, the time scale Δt is determined by the pressure p and the density ρ. Time scale and length scale may not be varied independently!

Table C.1: Physical properties of liquid aluminum and magnesium [123].

	ρ (Mg/m^3)	η (mPa· s)	ν (m^2/s)	σ (N/m)	v_{ca} (m/s)
Al	2.38	1.20	$0.50 \cdot 10^{-6}$	0.91	777
Mg	1.59	1.25	$0.79 \cdot 10^{-6}$	0.56	448

Table C.2: Numerical experiment Sim1:
Global parameters: $\Delta x = 10\ \mu$m, $\Delta t = 0.3 \cdot 10^{-6}$ s, $\Delta m = 2.4 \cdot 10^{-9}$g, $p = 9$ bar, $\rho = 2.38$ g/cm^3, $g = 0.$, $S^\star = 0.1$, $Q^\star = 10^{-5}$, $D^\star = 0.025$, expansion velocity $v = 0.005$ m/s, 200×200 cells, 75 bubble nuclei, n_{steps}: $8 \cdot 10^4$, periodic boundary conditions in x-direction.

Experiment	$\tau^\star$	$\sigma^\star$	$\nu^\star/\sigma^\star$	v_{ca} (m/s)	Ca
Sim1	0.55	0.03	0.55	61.7	$0.82 \cdot 10^{-3}$

Table C.3: Numerical experiments Sim2 - Sim6:
Global parameters: $\Delta x = 25\ \mu$m, $\Delta t = 2.25 \cdot 10^{-6}$ s, $\Delta m = 3.75 \cdot 10^{-8}$ g, $p = 1$ bar, $\rho = 2.38$ g/cm^3, $g = 0.$, $S^\star = 0.1$, $Q^\star = 10^{-5}$, $D^\star = 0.025$, $d^\star_\pi = 3\Delta x$, expansion velocity $v = 0.001$ m/s, 400×400 cells, 300 bubble nuclei, n_{steps}: $1.4 \cdot 10^5$, periodic boundary conditions in x-direction.

Experiment	$\tau^\star$	$\sigma^\star$	$\nu^\star/\sigma^\star$	v_{ca} (m/s)	Ca
Sim2	0.6	0.003	10.79	1.04	0.0097
Sim3	0.7	0.0018	36.00	0.31	0.0322
Sim4	2.5	0.003	171.37	0.066	0.1533
Sim5	5.5	0.0006	2699.99	0.0042	2.42
Sim6	1.7	0.00006	5555.55	0.002	4.98

Table C.4: Numerical experiments Sim7 - Sim8:
Global parameters: $\Delta x = 25\ \mu$m, $\Delta t = 2.25 \cdot 10^{-6}$ s, $p = 1$ bar, $\rho = 2.38$ g/cm^3, $g = 0.$, $S^\star = 0.1$, $Q^\star = 10^{-5}$, $D^\star = 0.025$, $d^\star_\pi = 3\Delta x$, 400 × 600 cells, n_{steps}: $5 \cdot 10^5$, periodic boundary conditions in x-direction.

Experiment	$\tau^\star$	$\sigma^\star$	$\nu^\star/\sigma^\star$	v_{ca} (m/s)	Ca	$\frac{\Pi_{\max}}{\sigma/\Delta x}$
Sim7	0.7	0.0018	36.00	0.31	0.0322	0.45
Sim8	0.7	0.0018	36.00	0.31	0.0322	1.35

Table C.5: Numerical experiments Sim9 - Sim10:
Global parameters: $\Delta x = 25\ \mu$m, $\Delta t = 2.25 \cdot 10^{-6}$ s, $p = 1$ bar, $\rho = 2.38$ g/cm^3, $g = 0.$, $S^\star = 0.1$, $Q^\star = 10^{-5}$, $D^\star = 0.025$, 400 × 500 cells, $n_0 = 25/\text{mm}^2$, periodic boundary conditions in x-direction.

Experiment	$\tau^\star$	$\sigma^\star$	$\nu^\star/\sigma^\star$	v_{ca} (m/s)	Ca	$\frac{\Pi_{\max}}{\sigma/\Delta x}$
Sim9	0.7	0.0018	36.00	0.31	0.0322	0.45
Sim10	5.5	0.0006	2699.99	0.0042	2.42	0.9

Table C.6: Numerical experiments Endo1 + Endo2:
Global parameters: $\Delta x = 2\ \mu$m, $\Delta t = 2.25 \cdot 10^{-6}$ s, $\Delta m = 1.9 \cdot 10^{-11}$ g, $p = 64$ mbar, $\rho = 2.38$ g/cm^3, $\Pi = 0.$, $g = 0.$, expansion velocity $v = 0.001$ m/s, 800 × 700 cells, $n_0 = 156/\text{mm}^2$, periodic boundary conditions in x-direction.

Experiment	$\tau^\star$	$\sigma^\star$	$\nu^\star/\sigma^\star$	v_{ca} (m/s)	Ca	V_{rel}
Endo1	0.65	0.01	65.00	0.014	0.071	0.34
Endo2	0.65	0.01	65.00	0.014	0.071	0.68

Symbols

Abbreviations

CA	Cellular automata
CFD	Computational fluid dynamics
IFM	Integral foam molding
HP-IFM	High pressure integral foam molding
JMA	Johnson-Mehl-Avrami
LBE	Lattice Boltzmann equation
LBM	Lattice Boltzmann method
LGA	Lattice gas automata
LP-IFM	Low pressure integral foam molding
NSE	Navier-Stokes equations
PM foam	Foam produced by powder compacts
μCT	Micro computer tomography

Mathematic Symbols

∂	Partial derivative
δ_{ij}	Kronecker delta function
∇	Nabla Operator
$\delta^{(3)}$	Delta function in three dimensions

Greek Symbols

δ	Mean material thickness
δ^{diff}	Diffusion length
Δm	Mass scale
Δn	Scale for the amount of substance
Δx	Lattice spacing = length scale
Δt	Time step = time scale
ϵ	Volume fraction of fluid within a cell
Γ	Liquid-gas interface
Π	Disjoining pressure
α	Spatial index
β	Spatial index
η	Dynamic viscosity
κ	Curvature
ν	Kinematic viscosity
ϕ	Phase ratio
ρ	Density
ρ_{rel}	Relative density
ρ_{F}	Foam density
σ	Surface tension
τ	Relaxation time
τ_{D}	Relaxation time for diffusion

Latin Symbols

B_{F}	Beam stiffness
Ca	Capillary number
D	Mean cell diameter
D	Gas diffusion constant
D_{p}	Particle diameter
E	Elastic modulus
I	Moment of Inertia
Q^{-1}	Internal friction
R	Gas constant
Re	Reynolds number
R_{Pl}	Curvature radius of the Plateau border
R_{p}	Particle radius
S	Sieverts' constant
T	Temperature
V_i	Volume of cell (bubble) i
V_{rel}	Relative particle volume fraction

Latin Symbols

c	Gas concentration
c_Π	Disjoining pressure constant
c_s	Speed of sound
d	Cell wall thickness
d_{crit}	Critical cell wall thickness
d_π	Range of the disjoining pressure
f	Phase fraction
$\boldsymbol{g}$	Gravitational acceleration
k_B	Boltzmann constant
m	Mass
m	Spacial dimension, 2D: $m = 2$, 3D: $m = 3$
n	Amount of substance
n	Evolution exponent
n_0	Nuclei density
n_{steps}	Total number of time steps
p	Pressure
p_0	Atmospheric or external pressure
p_i	Pressure of cell (bubble) i
p_{cast}	Casting pressure
p_{dwell}	Dwell pressure
t	Time
$\mathbf{v}$	Velocity
v_{ca}	Capillary velocity
$\mathbf{x}$	Spatial coordinate

Bibliography

[1] D.G.A.L. Aarts, M. Schmidt, and H.N.W. Lekkerkerker. *Direct visual observation of thermal capillary waves.* Science, **304**: 847–850, 2004.

[2] B.C. Allen, M.W. Mote, and A.M. Sabroff. US Patent 3,087,807, 1963.

[3] A. Andreasen, M.B. Sørensen, R. Burkarl, B. Møller, A.M. Molenbroek, S. Pedersen, J.W. Andreasen, M.M. Nielsen, and T.R. Jensen. *Interaction of hydrogen with an Mg-Al alloy.* Journal of Alloys and Compounds, **404–406**: 323–326, 2005.

[4] M. Arnold, C. Körner, and R.F. Singer. *PM aluminium foams: stabilizing mechanisms and optimization.* In J. Banhart et al., editor, Cellular Metals. Manufacture, Properties, Applications, pp. 71–76. Verlag MIT Publishing, Berlin 2003.

[5] M.F. Ashby, A. Evans, N.A. Fleck, L.J. Gibson, J.W. Hutchinson, and H.N.G. Wadley. *Metal Foams: A Design Guide.* Butterworth Heinemann, 2000.

[6] M. Avrami. *Kinetics of phase change I (General theory).* Journal of Chemical Physics, **7**: 1103–1112, 1939.

[7] N. Babcsán, D. Leitlmeier, and H.P. Degischer. *Foamability of particle reinforced aluminum melt.* Materialwissenschaft und Werkstofftechnik, **34**: 22–29, 2003.

[8] N. Babcsán, D. Leitlmeier, H.P. Degischer, and J. Banhart. *The role of oxidation in blowing particle-stabilized aluminium foams.* Advanced Engineering Materials, **6**: 421–428, 2004.

[9] N. Babcsán, D. Leitlmeier, H.P. Degischer, and H.J. Flankl. *Particle stabilized liquid-aluminum foams and the role of the oxide skin.* In J. Banhart et al., editor, Cellular Metals. Manufacture, Properties, Applications, pp. 101–106. Verlag MIT Publishing, Berlin 2003.

[10] N. Babcsán, I. Meszaros, and N. Hegman. *Thermal and electrical conductivity measurements on aluminum foams.* Materialwissenschaft und Werkstofftechnik, **34**: 391–394, 2003.

[11] J. Banhart. *Manufacture, characterization and application of cellular metals and metal foams.* Progress in Materials Science, **46**: 559–632, 2001.

[12] J. Banhart. *Manufacturing routes for metallic foams.* The Journal of the Minerals, Metals and Materials Society, **52**: 22–27, 2002.

[13] J. Banhart and J. Baumeister. *Deformation characteristics of metallic foams.* Journal of Materials Science, **33**: 1431–1440, 1998.

[14] J. Banhart, J. Baumeister, and M. Weber. *Damping properties of aluminium foams.* Materials Science and Engineering A, **205**: 221–228, 1996.

[15] J. Banhart, H. Stanzick, L. Helfen, and T. Baumbach. *Metal foam evolution studied by synchrotron radioscopy.* Applied Physics Letters, **78**: 1125–1128, 2001.

[16] J. Banhart, H. Stanzick, L. Helfen, T. Baumbach, and K. Nijhof. *Real-time observation of aluminium foam sandwich production by synchrotron radioscopy.* Advanced Engineering Materials, **3**: 407–411, 2001.

[17] J. Baumeister. German Patent, 4,018,360, 1990.

[18] G. Beall. *Injection moulded structural foam – a designer's perspective.* Kunststoffe, **84**: 62–65, 1994.

[19] R. Benzi, S. Succi, and M. Vergassola. *The lattice Boltzmann equation: theory and applications.* Physics Reports, **222**: 145–197, 1992.

[20] J. Bernsdorf, G. Brenner, and F. Durst. *Numerical analysis of the pressure drop in porous media flow with lattice Boltzmann (BGK) automata.* Computer Physics Communications, **129**: 247–255, 2000.

[21] A. Bhakta and E. Ruckenstein. *Decay of standing foams: drainage, coalescence and collapse.* Advances in Colloid and Interface Science, **70**: 1–124, 1997.

[22] P. Bhatnagar, E. Gross, and M. Krook. *A model for collision process in gases I: small amplitude processes in charged and neutral one-component system.* Physical Review E, **50**: 511–525, 1954.

[23] J. Bikar, J. Bruchon, T. Coupez, and B. Vergnes. *Numerical prediction of the foam structure of polymeric materials by direct 3D simulation of their expansion by chemical reaction based on a multidomain method.* Journal of Materials Science, **40**: 5875–5881, 2005.

[24] S.K. Bindal, G. Sethumadhavan, A.D. Nikolov, and D.T. Wasan. *Foaming mechanisms in surfactant free particle suspensions.* Materials, Interfaces, and Electrochemical Phenomena, **48**: 2307–2314, 2002.

[25] U. Binder, B. Fritsche, and W. Knott. *COMBAL – The new one step process for an economic aluminium foam production.* Advanced Engineering Materials, **8**: 778–780, 2006.

[26] B.P. Binks. *Particles as surfactants – similarities and differences.* Current Opinion in Colloid & Interface Science, **7**: 21–41, 2002.

[27] P. Boltenhagen and N. Pittet. *Structural transitions in crystalline foams.* Europhysics Letters, **41**: 571–576, 1998.

[28] F. Bolton and D. Weaire. *The effects of Plateau borders in the two-dimensional soap froth.* Philosophical Magazine B, **63**: 795–809, 1991.

[29] K. Boomsma, D. Poulikados, and Y. Ventikos. *Simulations of flow through open cell metal foams using an idealized periodic cell structure.* International Journal of Heat and Fluid Flow, **24**: 825–834, 2003.

[30] R.F. Brooks, A.T. Dinsdale, and P.N. Quested. *The measurement of viscosity of alloys – a review of methods, data and models.* Measurement Science and Technology, **16**: 354–362, 2005.

[31] J. Bruchon and R. Coupez. *Étude 3D de la formation d'une structure de mousse polymère par simulation de l´expansion anisotherme de bulles de gaz.* Mécanique & Industries, **4**: 331–338, 2003.

[32] M. Brunnbauer, C. Körner, and R.F. Singer. *Encasing of Aluminium Foam with Magnesium by Die Casting.* In J. Banhart et al., editor, Cellular Metals. Manufacture, Properties, Applications, pp. 187–190. Verlag MIT Publishing, Berlin 2003.

[33] E. Brunnhuber. *Praxis der Druckgussfertigung.* Schiele & Schöne, Berlin, 1991.

[34] L.N. Brush and S.H. Davis. *A new law of thinning in foam dynamics.* Journal of Fluid Mechanics, **534**: 227–236, 2005.

[35] M. Buchmann, S. Busch, A. Jaeger, and R. Klenz. *Nicht die Blasengröße ist entscheidend.* Kunststoffe, **10**: 216–220, 2005.

[36] J.M. Buick and C.A. Greated. *Gravity in a lattice Boltzmann model.* Physical Review E, **61**: 5307–5320, 2000.

[37] J.J. Chae. *Dynamics of foams with and without wall rupture.* Physical Review E, **55**: 598–610, 1997.

[38] H. Chen, C. Teixeira, and K. Molvig. *Realization of fluid boundary conditions via discrete Boltzmann dynamics.* International Journal of Modern Physics, **9**: 1281–1292, 1998.

[39] J.Y. Chen and Z. Fan. *Modelling the rheological behavior of semisolid metal slurries. Part 1 – Theory.* Materials Science and Technology, **18**: 237–242, 2002.

[40] S. Chen and G.D. Doolen. *Lattice Boltzmann method for fluid flows.* Annual Review of Fluid Mechanics, **30**: 329–364, 1998.

[41] X.-G. Chen and S. Engler. *Wasserstoff und Porosität in Aluminium-Silicium und Aluminium-Magnesium-Legierungen. Teil II: Aluminium-Magnesium-Legierungen und Diskussion über die Porenbildung.* Materialwissenschaft und Technik, **45**: 1225–1231, 1991.

[42] F. Chinesta, E. Cueto, P. Quintela, and J. Paredes. *Induced anisotropy in foams forming processes: modelling and simulation.* Journal of Materials Processing Technology, **155–156**: 1482–1488, 2004.

[43] B. Chopard and M. Droz. *Cellular Automata Modeling of Physical Systems.* Cambridge University Press, 1998.

[44] B. Chopard and O. Luthi. *Lattice Boltzmann computations and applications to physics.* Theoretical Computer Science, **217**: 115–130, 1999.

[45] M. Öchsner, M. Thies, M. Arnold, C. Körner, and R.F. Singer. *Simulation of metal foam formation with the lattice Boltzmann method.* In R.F. Singer, C. Körner, V. Altstädt, and H. Münstedt, editors, Cellular Metals and Polymers, pp. 45–48. Trans Tech Publications, 2005.

[46] X.L. Chu, A.D. Nikolov, and D.T. Wasan. *Effects of particle size and polydispersity on the depletion and structural forces in colloidal dispersions.* Langmuir, **12**: 5004–5010, 1994.

[47] X.L. Chu, A.D. Nikolov, and D.T. Wasan. *Monte Carlo simulation of inlayer struture formation in thin liquid films.* Langmuir, **10**: 4403–4408, 1994.

[48] X.L. Chu, A.D. Nikolov, and D.T. Wasan. *Thin liquid film structure and stability: the role of depletion and surface-induced structural forces.* Journal of Chemical Physics, **103**: 6653–6661, 1995.

[49] D. Cohen. *Ray Tracing*, chapter Voxel traversal along a 3D line, pp. 366–369. Academic Press, 1994.

[50] J.S. Colton. *Nucleation of microcellular foam: theory and practice.* Polymer Engineering and Science, **27**: 500–503, 1987.

[51] H.-P. Degischer and B. Kriszt, editors. *Handbook of Cellular Metals.* WILEY-VCH, 2002.

[52] F. Delannay, L. Froyen, and A. Deruyttere. *Review: the wetting of solids by molten metals and its relation to the preparation of metal-matrix composites.* Journal of Materials Science, **22**: 1–16, 1987.

[53] W. Deqing, M. Xiangjun, X. Weiwei, and S. Ziyuan. *Effect of processing parameters on cell structure of an aluminum foam.* Materials Science and Engineering A, **420**: 235–239, 2006.

[54] V.S. Deshpande and N.A. Fleck. *High strain rate compressive behavior of aluminum alloy foams.* International Journal of Impact Engineering, **24**: 277–298, 2000.

[55] A. Dippenaar. *The destabilization of froth by solids. I. The mechanism of film rupture.* International Journal of Mineral Processing, **9**: 1–14, 1982.

[56] W. Drenckhan, S.J. Cox, G. Delaney, H. Holste, D. Weaire, and N. Kern. *Rheology of ordered foams – on the way to discrete microfluids.* Colloids and Surfaces A: Physicochemical Engineering Aspects, **263**: 52–64, 2005.

[57] D.J. Durian. *Foam mechanics at the bubble scale.* Physical Review E, **75**: 4780–4783, 1995.

[58] D.J. Durian. *Bubble scale model of foam mechanics: melting, nonlinear behavior, and avalanches.* Physical Review E, **55**: 1739–1751, 1997.

[59] H. Eckardt. *Thermoplastic structural foam – a well known and new process.* In Blowing Agents and Foaming Processes 2003 conference, pp. 65–76, München, 2003.

[60] S.F. Edwards and K.D. Pithia. *A model for the formation of foams.* Physica A, **215**: 270–276, 1995.

[61] P. Egger, M. Fischer, H. Kirschling, and A.K. Bledzki. *Serienfeste Vielseitigkeit beim MuCell-Spritzgießen (1).* Kunststoffe, **12**: 66–70, 2005.

[62] P. Egger, M. Fischer, H. Kirschling, and A.K. Bledzki. *Serienfeste Vielseitigkeit beim MuCell-Spritzgießen. Ein Statusbericht für die Praxis (2).* Kunststoffe, **1**: 72–76, 2006.

[63] E.A. Egli. *Design properties of structural foam.* Journal of Cellular Plastics, pp. 245–249, 1972.

[64] W. Eichenauer and J. Makropoulus. *Messung des Diffusionkoefizienten von Wasserstoff in flüssigem Aluminium.* Zeitschrift für Metallkunde, **65**: 649–652, 1974.

[65] S. Engler. *Das Volumendefizit bei der Erstarrung metallischer Schmelzen.* Zeitschrift für Metallkunde, **56**: 327–335, 1965.

[66] S. Engler. *Zur Erstarrungsmorphologie von Aluminium-Gußwerkstoffen. Teil I: Grundlagen und Versuchsergebnisse an Reinst- und Reinaluminium sowie an AlMg- und AlZnMg-Legierungen.* Aluminium, **45**: 673–678, 1969.

[67] A.G. Evans, J.W. Hutchinson, and M.F. Ashby. *Multifunctionality of cellular metal systems.* Progress in Materials Science, **43**: 171–221, 1999.

[68] Z. Fan and J.Y. Chen. *Modelling of rheological behavior of semisolid metal slurries. Part 2 – Steady state behaviour.* Materials Science and Technology, **18**: 243–249, 2002.

[69] J.F. Fernández and C.R. Sánchez. *Rate determining step in the absorption and desorption of hydrogen by magnesium.* Journal of Alloys and Compounds, **340**: 189–198, 2002.

[70] J.F. Fernández and C.R. Sánchez. *Simultaneous TDC-DSC measurement in magnesium hydride.* Journal of Alloys and Compounds, **356–357**: 348–352, 2003.

[71] M. Fischer. *Strukturschaum-Spritzgießen zur Herstellung dünnwandiger Teile.* Maschinenmarkt, **48**: 24–28, 2000.

[72] M.A. Fortes. *The average rate of growth of individual cells in a cellular structure: effect of the number of topological elements.* Journal of Materials Science, **21**: 2509–2513, 1986.

[73] W. Fragner and H. Kaufmann. *Legierungsgerechte Auswahl von Gießparametern.* In H. Kaufmann and P.J. Uggowitzer, editors, 3. Ranshofener Leichtmetalltage: Vom Werkstoff zum Bauteilsystem, pp. 309–316. Ranshofen: LKR-Verlag, 2004.

[74] U. Frisch, D. d´Humières, B. Hasslacher, P. Lallemand, Y. Pomeau, and J.-P. Rivet. *Lattice gas hydrodynamics in two and three dimensions.* Complex Systems, **1**: 649–707, 1987.

[75] G.C. Frye and J.C. Berg. *Antifoam action of solid particles.* Journal of Colloid and Interface Science, **127**: 222–238, 1989.

[76] K. Fuchizaki and K. Kawasaki. *Time evolution of three-dimensional cellular systems: computer modeling based on vertex type models.* Physica A, **221**: 202–215, 1995.

[77] B.S. Gardiner, B.Z. Dlugogorski, and G.J. Jameson. *The steady shear of three-dimensional wet polydisperse foams.* Journal of Non-Newtonian Fluid Mechanics, **92**: 151–166, 2000.

[78] V. Gergely. *Melt Route Processing for Production of Metallic Foams.* PhD Thesis, University of Cambridge, 2000.

[79] V. Gergely and T.W. Clyne. *The FORMGRIP Process: Foaming of Reinforced Metals by Gas Release in Precursor.* Advanced Engineering Materials, **2**: 175–178, 2000.

[80] V. Gergely and T.W. Clyne. *Drainage in standing liquid metal foams: modeling and experimental observations.* Acta Materialia, **52**: 3047–3058, 2004.

[81] V. Gergely, D.C. Curran, and T.W. Clyne. *The FOAMCARP process: foaming of aluminium MMCs by the chalk–aluminium reaction in precursors.* Composites Science and Technology, **63**: 2301–2310, 2003.

[82] V. Gergely, D.C. Curran, and T.W. Clyne. *Drainage of metallic foams – modelling and experimental observations.* In J. Banhart et al., editor, Cellular Metals. Manufacture, Properties, Applications, pp. 105–110. Verlag MIT Publishing, Bremen 2001.

[83] M. Gerhardt and H. Schuster. *Das digitale Universum.* Vieweg, 1995.

[84] L.J. Gibson. *Mechanical behavior of metallic foams.* Annual Review of Materials Science, **30**: 191–227, 2000.

[85] L.J. Gibson and M.F. Ashby. *Cellular solids.* Cambridge University Press, 1997.

[86] I. Ginzburg and K. Steiner. *A free-surface lattice Boltzmann method for modelling the filling of expanding cavities by Bingham fluids.* Philosophical Transactions of the Royal Society of London A, **360**: 453–466, 2002.

[87] I. Ginzburg and K. Steiner. *Lattice Boltzmann model for free-surface flow and its application to filling process in casting.* Journal of Computational Physics, **185**: 61–99, 2003.

[88] J. Göken and W. Riehemann. *Thermoelastic damping of the low density metals AZ91 and DISPAL.* Materials Science and Engineering A, **324**: 134–140, 2002.

[89] J. Glazier and D. Weaire. *The kinetics of cellular patterns.* Journal of Physics: Condensed Matter, **4**: 1867–1894, 1992.

[90] Goldschmidt. *Sicherheitsdatenblatt Tego Magnan, Version 2.1.* 2002.

[91] I.S. Golovin and H.-R. Sinning. *Damping of some aluminum foams at low amplitudes.* Solid State Phenomena, **89**: 261–266, 2003.

[92] I.S. Golovin and H.-R. Sinning. *Internal friction in metallic foams and some related cellular structures.* Materials Science and Engineering A, **370**: 504–511, 2004.

[93] I.S. Golovin, H.-R. Sinning, I.K. Arhipov, S.A. Golovin, and M. Bram. *Damping in some cellular metallic materials due to microplasticity.* Materials Science and Engineering A, **370**: 531–536, 2004.

[94] I.S. Golovin, H.-R. Sinning, J. Göken, and W. Riehemann. *Fatique-related damping in some cellular metallic materials.* Materials Science and Engineering A, **370**: 537–541, 2004.

[95] P. Grassia, S.J. Neethling, C. Cervantes, and H.T. Lee. *The growth, drainage and bursting of foams.* Colloids and Surfaces A: Physicochemical Engineering Aspects, **274**: 110–124, 2006.

[96] A. Gray. *Modern Differential Geometry of Curves and Surfaces with Mathematica.* CRC Press, 1997.

[97] M.C. Gui, D.B. Wang J.J. Wu, G.J. Yuan, and C.G. Li. *Deformation and damping behaviors of foamed Al-Si-SiC$_p$ composite.* Materials Science and Engineering A, **286**: 282–288, 2000.

[98] A. Gunstensen and D.H. Rothman. *Microscopic modeling of immiscible fluids in three dimensions by a lattice Boltzmann method.* Europhysics Letters, **18**: 157–161, 1992.

[99] A. Gunstensen, D.H. Rothman, S. Zaleski, and G. Zanetti. *Lattice Boltzmann Model of immiscible fluids.* Physical Review A, **43**: 4320–4327, 1991.

[100] F. Han, Z. Zhu, and J. Gao. *Compressive deformation and energy absorbing characteristic of foamed aluminum.* Metallurgical and Materials Transactions A, **29A**: 2497–2502, 1998.

[101] F. Han, Z. Zhu, C. Liu, and J. Gao. *Damping behavior of foamed aluminum.* Metallurgical and Materials Transactions A, **30A**: 771–776, 1999.

[102] W. Hanselmann and E. Windhab. *Foam flow in pipes.* Applied Rheology, pp. 253–260, 1996.

[103] A.G. Hanssen, M. Langseth, and O.S. Hopperstad. *Optimum design for energy absorption of square aluminium columns with aluminium foam filler.* International Journal of Mechanical Sciences, **43**: 153–176, 2001.

[104] Y. Hashimoto and H. Ohashi. *Droplet dynamics using the lattice-gas method.* International Journal of Modern Physics C, **8**: 977–983, 1997.

[105] S. Hasse. *Guss- und Gefügefehler.* Schiele & Schön. Berlin, 2002.

[106] X. He and L.-S. Luo. *A priori derivation of the lattice Boltzmann equation.* Physical Review E, **55**: R6333–6336, 1997.

[107] X. He and L.-S. Luo. *Theory of the lattice Boltzmann method: From the Boltzmann equation to the lattice Boltzmann equation.* Physical Review E, **56**: 6811–6817, 1997.

[108] F. Heinrich, C. Körner, and R.F. Singer. *Herstellung flächiger Leichtbauteile durch Umgießen von Aluminium-Schäumen im Druckgießverfahren.* In H.-P. Degischer, editor, Metallschäume, pp. 428–431. Wiley-VCH, Wien 2000.

[109] R. Höhler, S. Cohen-Addad, and A. Asnacios. *Rheological memory effect in aqueous foam.* Europhysics Letters, **48**: 93–98, 1999.

[110] F. Higuera and J. Jimenez. *Boltzmann approach to lattice gas simulations.* Europhysics Letters, **9**: 663–668, 1989.

[111] N.C. Hilyard. *Mechanics of cellular plastics.* Applied Science Publishers LTD, London, 1982.

[112] J.A. Hines. *Determination of interfacial heat-transfer boundary conditions in an aluminum low-pressure permanent mold test casting.* Metallurgical and Materials Transactions B, **35**: 299–311, 2004.

[113] M. Hirschmann, C. Körner, and R.F. Singer. *Magnesiumintegralschäume.* Gießereiforschung, **57**: 8–13, 2005.

[114] M. Hirschmann, C. Körner, and R.F. Singer. *Integral foam molding – a new process for foamed magnesium castings.* Materials Science Forum, **539–543**: 1827–1832, 2007.

[115] M. Hirschmann, H. Wiehler, C. Körner, and R.F. Singer. *Light metal integral foam – a promising material for structural applications.* In G. Stephani and B. Kieback, editors, Cellular Metals for Structural and Functional Applications, pp. 147–154. Fraunhofer IRB Verlag, 2005.

[116] H. Hoballah, R. Höhler, and S. Cohen-Added. *Time evolution of the elastic properties of aqueous foam.* Journal de Physique II, **7**: 1215–1224, 1997.

[117] S. Hou, X. Shan, Q. Zou, G.D. Doolen, and W.E. Soll. *Evalution of two lattice Boltzmann models for multiphase flows.* Journal of Computational Physics, **138**: 695–713, 1997.

[118] S. Hutzler, S.J. Cox, and G. Wang. *Foam drainage in two dimensions.* Colloids and Surfaces A: Physicochemical Engineering Aspects, **263**: 178–183, 2005.

[119] S. Hutzler, D. Weaire, and R. Crawford. *Convective instability in foam drainage.* Europhysics Letters, **41**: 461–465, 1998.

[120] S. Hutzler, D. Weaire, and S. Shah. *Bubble sorting in a foam under forced drainage.* Philosophical Magazine Letters, **80**: 41–48, 2000.

[121] A. Ibrahim, C. Körner, and R.F. Singer. *Investigation of micropore creation on the PM-foam morphology.* In R. F. Singer, C. Körner, V. Altstädt, and H. Münstedt, editors, Cellular Metals and Polymers, pp. 61–64. Trans Tech Publications, 2005.

[122] A.H. Ibrahim. *Effect of material and processing parameters on the morphology of aluminium foams produced by the PM route.* PhD Thesis, University of Erlangen-Nürnberg, 2005.

[123] T. Iida and R.I.L. Guthrie. *The Physical Properties of Liquid Metals.* Clarendon Press · Oxford, 1988.

[124] S.W. Ip, Y. Wang, and J.M. Toguri. *Aluminum foam stabilization by solid particles.* Canadian Metallurgical Quarterly, **38**: 81–92, 1999.

[125] H.M. Jaeger and S.R. Nagel. *Physics of Granular States.* Science, **255**: 1524, 1992.

[126] Y. Jiang, P.J. Swart, A. Saxena, M. Asipauskas, and J.A. Glazier. *Hysteresis and avalanches in two-dimensional foam rheology simulations.* Physical Review E, **59**: 5819–5832, 1999.

[127] W. Jiejun, L. Chenggong, W. Dianbin, and G. Manchang. *Damping and sound absorption properties of particle reinforced Al matrix composite foams.* Composites Science and Technology, **63**: 569–574, 2003.

[128] F. Johannaber and W. Michaeli. *Handbuch Spritzgießen.* Hanser, München, 2004.

[129] W.A. Johnson and R.F. Mehl. *Reaction kinetics in processes of nucleation and growth.* Transactions of the American Institute of Minerals (Metall.) Engs., **135**: 416–441, 1939.

[130] C. Kammer. *Magnesium-Taschenbuch.* Aluminium, 2000.

[131] G. Kaptay. *Classification of interfacial forces acting on solid particles in technologies of advanced metallic materials.* In International Conference Advanced Metallic Materials, 5–7 November, Slovakia, pp. 135–140, 2003.

[132] G. Kaptay. *Interfacial criteria for stabilization of liquid foams by solid particles.* Colloids and Surfaces A: Physicochemical Engineering Aspects, **230**: 67–80, 2004.

[133] G. Kaptay. *On the equation of the maximum capillary pressure induced by solid particles to stabilize emulsions and foams and on the emulsion stability diagrams.* Colloids and Surfaces A: Physicochemical Engineering Aspects, **282–283**: 387–401, 2006.

[134] G. Kaptay. *An improved interfacial model of stabilization of metallic foams by solid particles.* In J. Banhart et al., editor, Cellular Metals. Manufacture, Properties, Applications, pp. 107–112. Verlag MIT Publishing, Berlin 2003.

[135] A.T.W. Kempen. *Solid State Phase Transformation Kinetics.* PhD Thesis, Stuttgart, 2001.

[136] J.P. Kermode and D. Weaire. *2D-froth: A program for the investigation of 2-dimensional froths.* Computer Physics Communications, **60**: 75–109, 1990.

[137] S.A. Koehler, S. Hilgenfeldt, and H.A. Stone. *Liquid flow through aqueous foams: the node-dominated foam drainage equation.* Physical Review Letters, **82**: 4232–4235, 1999.

[138] S.A. Koehler, S. Hilgenfeldt, and H.A. Stone. *Foam drainage on the microscale I. Modeling flow through single Plateau borders.* Journal of Colloid and Interface Science, **276**: 420–438, 2004.

[139] J. Koke and M. Modigell. *Flow behaviour of semi-solid metal alloys.* Journal of Non-Newtonian Fluid Mechanics, **112**: 141–160, 2003.

[140] K. Kono, T. Ishizuka, H. Tsuda, and A. Kurosawa. *Application of lattice Boltzmann model to multiphase flows with phase transitions.* Computer Physics Communications, **129**: 110–120, 2000.

[141] D. Kothe, D. Juric, K. Lam, and B. Lally. *Numerical recipes for mold filling simulations.* Proceedings of the Eighth International Conference on Modeling of Casting, Welding and Advanced Solidification Processes, 1998.

[142] A.M. Kraynik. *Foam flows.* Annual Review of Fluid Mechanics, **20**: 325–357, 1988.

[143] A.M. Kraynik and D.A. Reinelt. *Linear elastic behavior of dry soap foams.* Journal of Colloid and Interface Science, **181**: 511–520, 1996.

[144] R. Kretz and E. Wolfsgruber. *Bedingungen für das Eingießen von Aluminium-Schaumkernen in Leichtmetallgussteile.* Gießerei, **90**: 50–59, 2003.

[145] I.M. Krieger and T. Dougherty. *A mechanism for non-newtonian flow in suspensions of rigid spheres.* Transactions of the Society of Rheology, **3**: 137–152, 1959.

[146] C. Körner, M. Arnold, and R.F. Singer. *Metal foam stabilization by oxide network particles.* Materials Science and Engineering, **A 396**: 28–40, 2005.

[147] C. Körner, F. Berger, M. Arnold, C. Stadelmann, and R.F. Singer. *Influence of processing conditions on morphology of metal foams produced from metal powder.* Materials Science and Technology, **16**: 781–784, 2000.

[148] C. Körner, F. Heinrich, and R.F. Singer. *Encasing by Casting.* In H.-P. Degischer and B. Kriszt, editors, Handbook of Cellular Metals, pp. 103–113. Wiley-VCH, 2002.

[149] C. Körner, M. Hirschmann, V. Braeutigam, and R.F. Singer. *Endogenous particle stabilization during magnesium integral foam production.* Advanced Engineering Materials, **6**: 385–390, 2004.

[150] C. Körner, M. Hirschmann, M. Lamm, and R.F. Singer. *Magnesium integral foams.* In J. Banhart et al., editor, Cellular Metals. Manufacture, Properties, Applications, pp. 209–214. Verlag MIT Publishing, Berlin 2003.

[151] C. Körner, M. Hirschmann, and R.F. Singer. DE 103 25 819 B4, 2003.

[152] C. Körner, M. Hirschmann, and H. Wiehler. *Integral foam moulding of light metals.* In H. Nakajima and N. Kanetake, editors, Porous Metals and Metals Foaming Technologies, pp. 279–284. The Japan Institute of Metals, 2005.

[153] C. Körner, M. Hirschmann, and H. Wiehler. *Integral foam moulding of light metals.* Materials Transactions, **47**: 2188–2194, 2006.

[154] C. Körner, T. Pohl, U. Rüde, N. Thürey, and T. Hofmann. *FreeWIHR: lattice Boltzmann methods with free surfaces and their applications in material technology.* In A. Bode and F. Durst, editors, High Performance Computing in Science and Engineering, pp. 225–236. Springer, Garching 2004.

[155] C. Körner, T. Pohl, U. Rüde, N. Thürey, and T. Zeiser. *Parallel Lattice Boltzmann Methods for CFD Applications.* In Numerical Solution of Partial Differential Equations on Parallel Computers, A.M. Bruaset, A. Tveito, editors, volume 51 of Lecture Notes in Computational Science and Engineering, pp. 437–466. Springer, 2006.

[156] C. Körner and R.F. Singer. *Processing of metal foams – challenges and opportunities.* Advanced Engineering Materials, **2**: 159–165, 2000.

[157] C. Körner and R.F. Singer. *Foaming processes for aluminum.* In H.P. Degischer and B. Kriszt, editors, Handbook of Cellular Metals, pp. 9–14. Wiley-VCH, 2002.

[158] C. Körner and R.F. Singer. *The physics of foaming: structure formation and stability.* In H. P. Degischer and B. Kriszt, editors, Handbook of Cellular Metals, pp. 33–42. Wiley-VCH, 2002.

[159] C. Körner and R.F. Singer. *Numerical simulation of foam formation and evolution with modified cellular automata.* In J. Banhart, M.F. Ashby, and N.A. Fleck, editors, Metal foams and Porous Metal Structures, pp. 91–96. Verlag MIT Publishing, Bremen 1999.

[160] C. Körner and M. Thies. *Foaming of light metals: mechanisms and numerical modeling.* In P. Vincenzini, editor, Proceeding of the 3rd International Conference "Computational Modeling and Simulation of Materials", volume 43, pp. 471–478. Advances in Science and Technology, Acireale 2004.

[161] C. Körner, M. Thies, M. Arnold, and R.F. Singer. *Modeling of metal foaming by in-situ gas formation.* In J. Banhart et al., editor, Cellular Metals. Manufacture, Properties, Applications, pp. 93–98. Verlag MIT Publishing, Bremen 2001.

[162] C. Körner, M. Thies, T. Hofmann, N. Thürey, and U. Rüde. *Lattice Boltzmann model for free surface flow for modeling foaming.* Journal of Statistical Physics, **121**: 179–196, 2005.

[163] C. Körner, M. Thies, and R.F. Singer. *Modeling of metal foaming with lattice Boltzmann automata.* Advanced Engineering Materials, **4**: 765–769, 2002.

[164] W. Kurz and D.J. Fischer. *Fundamentals of Solidification.* Trans Tech Publications, Swiss, 3rd edition, 1992.

[165] A. Ladd. *Numerical simulations of particulate suspensions via a discretized Boltzmann equation. Part 1.* Journal of Fluid Mechanics, **271**: 285–309, 1994.

[166] A. Ladd. *Numerical simulations of particulate suspensions via a discretized Boltzmann equation. Part 2.* Journal of Fluid Mechanics, **271**: 311–339, 1994.

[167] A. Ladd and R. Verberg. *Lattice Boltzmann simulations of particle–fluid suspensions.* Journal of Statistical Physics, **104**: 1191–1251, 2001.

[168] L.D. Landau and E.M. Lifschitz. *Hydrodynamik.* Akademie-Verlag · Berlin, 1974.

[169] H. Lange. *Schäume und ihre Stabilität.* VDI-Berichte, **182**: 71–77, 1972.

[170] R.G. Larson. *The elastic stress in film fluids.* Journal of Rheology, **41**: 365–372, 1997.

[171] C. Liu, Z. Zhu, F. Han, and J. Banhart. *Internal friction of foamed aluminium in the range of acoustic frequencies.* Journal of Materials Science, **33**: 1769–1775, 1998.

[172] C. Liu, Z. Zhu, F. Han, and J. Banhart. *Low-frequency internal friction of foamed Al.* Philosophical Magazine A, **78**: 1329–1337, 1998.

[173] A. Lohmüller, M. Scharrer, M. Hilbinger, R. Jenning, M. Hartmann, and R.F. Singer. *Thixospritzgießen von Magnesiumlegierungen. Thixomoulding of magnesium alloys.* Gießereiforschung, **57**: 2–9, 2005.

[174] L.-S. Luo. *Theory of the lattice Boltzmann method: lattice Boltzmann models for non-ideal gases.* Physical Review E, **62**: 4982–4996, 2000.

[175] P. Lutze and J. Ruge. *Wasserstoff in Aluminium und seinen Legierungen.* Metall, **44**: 741–748, 1990.

[176] L. Ma and Z. Song. *Cellular structure control of aluminium foams during foaming process of aluminium melt.* Scripta Materialia, **39**: 1523–1528, 1998.

[177] A. Mantey, J. Moll, D. Kunz, J.K.W. Sandler, and V. Altstädt. *Foam injection-molding of low density polymer components.* In R. F. Singer, C. Körner, V. Altstädt, and H. Münstedt, editors, Cellular Metals and Polymers, pp. 253–256. Trans Tech Publications, 2005.

[178] G.R. McNamara and G. Zanetti. *Use of the Boltzmann equation to simulate lattice-gas automata.* Physical Review Letters, **61**: 2332–2335, 1988.

[179] R. Mei, D. Yu, W. Shyy, and L.-S. Luo. *Force evaluation in the lattice Boltzmann method involving curved geometry.* Physical Review E, **65**: 041203/1–14, 2002.

[180] E.J. Mittemeijer. *Analysis of the kinetics of phase transformations.* Journal of Materials Science, **27**: 3977–3987, 1992.

[181] T. Miyoshi, M. Itoh, S. Akiyama, and A. Kitahara. *ALPORAS aluminum foam: production process, properties, and applications.* Advanced Engineering Materials, **2**: 179–183, 2000.

[182] N. Müller and G.W. Ehrenstein. *Evaluation and modeling of injection-molded rigid polypropylene integral foam.* Journal of Cellular Plastics, **40**: 45–59, 2004.

[183] Norbert Müller. *Spritzgegossene Integralschaumstrukturen mit ausgeprägter Dichtereduktion.* PhD- Thesis, University of Erlangen-Nürnberg, 2006.

[184] W. Müller and J.-M. de Meglio. *Avalanches in draining foams.* Journal of Condensed Matter, **11**: L209–L215, 1999.

[185] W. Müller and L. Starke. *Modelle zur Berechnung des Elastizitätsmoduls und der Biegesteifigkeit von thermoplastischen Strukturschaumstoffen.* Plaste und Kautschuk, **31**: 348–351, 1984.

[186] C. Monnereau and M. Vignes-Adler. *Dynamics of 3D real foam coarsening.* Physicsl Review Letters, **80**: 5228–5231, 1998.

[187] K. Mörwald. *Schaumspritzgießen.* Plastverarbeiter, **28**: 305–183, 1977.

[188] S.J. Neething and J.J. Cilliers. *Modelling flotation froths.* International Journal of Mineral Processing, **72**: 267–287, 2003.

[189] R. Neugebauer and T. Hipke. *Dynamisches Verhalten geschäumter Bauteile.* Materialwissenschaft und Werkstofftechnik, **31**: 515–518, 2000.

[190] B. Niedermann, W. Knott, M. Recksik, and A. Weier. EP 1356 131 B1, 2002.

[191] A.D. Nikolov and D.T. Wasan. *Dispersion stability due to structural contributions to the particle interaction as probed by thin liquid film dynamics.* Langmuir, **8**: 2985–2994, 1992.

[192] T. Okuzono and K. Kawasaki. *Intermittent flow behavior of random foams: a computer experiment of foam rheology.* Physical Review E, **51**: 1246–1253, 1995.

[193] O.B. Olurin, M. Arnold, C. Körner, and R.F. Singer. *The investigation of morphometric parameters of aluminium foams using micro-computed tomography.* Materials Science and Engineering A, **328**: 334–343, 2002.

[194] E. Ovrelid, P. Bakke, and T.A. Engh. *Solubility of hydrogen in magnesium alloys.* In International Symposium on Light Metals, pp. 141–154. Canadian Institute of Minining, Metallurgy and Petroleum, 1997.

[195] C. Pan, J.F. Prins, and C.T. Miller. *A high-performance lattice Boltzmann implementation to model flow in porous media.* Computer Physics Communications, **158**: 89–105, 2004.

[196] S.-H. Park and B.-Y. Hur. *A study on the viscosity and surface tension in molten Al and the effect of additional elements.* In J. Banhart et al., editor, Cellular Metals. Manufacture, Properties, Applications, pp. 123–128. Verlag MIT Publishing, Berlin 2003.

[197] M.S. Plesset and A. Prosperetti. *Bubble dynamics and cavitation.* Annual Reviews of Fluid Dynamics, **9**: 145–185, 1977.

[198] C. Pozrikidis. *Expansion of a two-dimensional foam.* Engineering Analysis with Boundary Elements, **26**: 495–504, 2002.

[199] R.J. Pugh. *Foaming, foam films, antifoaming and defoaming.* Advances in Colloid Interface Science, **64**: 67–142, 1996.

[200] Y.H. Qian, D. d´Humières, and P. Lallemand. *Lattice BGK models for Navier-Stokes equation.* Europhysics Letters, **17**: 479–484, 1992.

[201] Lord Rayleigh. *On the pressure developed in a liquid during the collapse of a spherical cavity.* Philosophical Magazine, **34**: 94–98, 1917.

[202] D.H. Rothman and S. Zaleski. *Lattice-gas Cellular Automata: Simple Models of Complex Hydrodynamics.* Cambridge University Press, 1997.

[203] L. Salvo, P. Cloetens, E. Maire, S. Zabler, J.J. Blandin, J.Y. Buffière, W. Ludwig, E. Boller, D. Bellet, and C. Josserond. *X-ray micro-tomography an attractive characterization technique in materials science.* Nuclear Instruments and Methods in Physics Research B, **200**: 273–286, 2003.

[204] M.V.C. Sastri, B. Viswanathan, and S.S. Murthy, editors. *Metal Hydrides.* Narosa Publishing House, 1998.

[205] F. Sauerwald. *Über die Löslichkeit und das Verhalten des Wasserstoffs in flüssigem und festem Magnesium und in Magnesiumlegierungen.* Zeitschrift für anorganische Chemie, **258**: 27–32, 1949.

[206] R. Scardovelli and S. Zaleski. *Direct numerical simulation of free-surface and interfacial flow.* Annual Review of Fluid Mechanics, **31**: 567–603, 1999.

[207] F. Schüth, B. Bogdanovic, and M. Felderhoff. *Light metal hydrides and complex hydrides for hydrogen storage.* Chemical Communication, pp. 2249–2258, 2004.

[208] L.W. Schwartz and R.V. Roy. *A mathematical model for an expanding foam.* Journal of Colloid and Interface Science, **264**: 237–249, 2003.

[209] H.-W. Seeliger. *Aluminium foam sandwich (AFS) ready for market introduction.* Advanced Engineering Materials, **6**: 448–451, 2004.

[210] S. Semerdjiev. *Introduction to structural foam.* Society of Plastics Engineers, Inc., 1982.

[211] G.N. Sethumadhavan, S. Bindal, A.D. Nikolov, and D.T. Wasan. *Stability of liquid films containing polydisperse particles.* Colloids and Surfaces, **204**: 51–62, 2002.

[212] G.N. Sethumadhavan, A.D. Nikolov, and D.T. Wasan. *Stability of liquid films containing monodisperse colloidal particles.* Journal of Colloid and Interface Science, **240**: 105–112, 2001.

[213] R.F. Singer and C. Körner. DE 100 45 494 C2, 2000.

[214] H. Stanzick, M. Wichmann, J. Weise, L. Helfen, T. Baumbach, and J. Banhart. *Process control in aluminum foam production using real-time X-ray radioscopy.* Advanced Engineering Materials, **4**: 814–823, 2002.

[215] J. Stavans. *The evolution of cellular structures.* Reports on Progress in Physics, **56**: 733–789, 1993.

[216] S. Stein and U. Dahn. *Wirtschaftlicher Leichtbau durch thermoplastische Schäume am Beispiel eines Instrumententafelträgers.* In O. Huber and M. Bicker, editors, Landshuter Leichtbau-Colloquium, pp. 87–97. LC-Verlag, 2005.

[217] G. Steinbichler, J. Kragl, D. Pierick, and K. Jacobsen. *Spritzgießen von Strukturschaum.* KU Kunststoffe, **89**: 50–54, 1999.

[218] H.A. Stone, S.A. Koehler, S. Hilgenfeldt, and M. Durand. *Perspectives on foam drainage and the influence of interfacial rheology.* Journal of Physics: Condensed Matter, **15**: S283–S290, 2003.

[219] S. Succi. *The Lattice Boltzmann Equation for Fluid Dynamics and Beyond.* Oxford Science, 2000.

[220] N.P. Suh. *Impact of microcellular plastics on industrial practice and academic research.* In Macromolecular Symposium, pp. 187–201. WILEY-VCH, Weinheim, 2003.

[221] Q. Sun and S. Hutzler. *Studying localized bubble rearrangements in 2D liquid foams using a hybrid lattice gas model.* Colloids and Surfaces A: Physicochemical Engineering Aspects, **263**: 27–32, 2005.

[222] N. Takada, M. Misawa, A. Tomiyama, and S. Fujiwara. *Numerical simulation of two- and three-dimensional two-phase fluid motion by lattice Boltzmann method.* Computer Physics Communications, **129**: 233–246, 2000.

[223] M. Thies. Lattice Boltzmann modeling with free surfaces applied to in-situ gas generated foam formation. PhD Thesis, University of Erlangen-Nürnberg, 2005.

[224] N. Thürey, T. Pohl, U. Rüde, M. Öchsner, and C. Körner. *Optimization and stabilization of LBM free surface flow simulations using adaptive parameterization.* Computers & Fluids, **35**: 934–939, 2006.

[225] Verein Deutscher Ingenieure VDI, editor. *Spritzgießen von Strukturschaumformteilen.* VDI-Verlag GmbH Düsseldorf, 1976.

[226] G. Verbist, D. Weaire, and A.M. Kraynik. *The foam drainage equation.* Journal of Physics: Condensed Matter, **8**: 3715–3731, 1996.

[227] T. Wübben, S. Odenbach, and J. Banhart. *Study of lead foam under microgravity.* In J. Banhart et al., editor, Cellular Metals. Manufacture, Properties, Applications, pp. 83–88. Verlag MIT Publishing, Bremen 2001.

[228] T. Wübben, H. Stanzick, J. Banhart, and S. Odenbach. *Stability of metallic foams studied under microgravity.* Journal of Physics: Condensed Matter, **15**: 427–433, 2003.

[229] D. Weaire. *Foam physics.* Advanced Engineering Materials, **4**: 723–725, 2002.

[230] D. Weaire. *Foam physics.* In J. Banhart et al., editor, Cellular Metals. Manufacture, Properties, Applications, pp. 63–68. Verlag MIT Publishing, Bremen 2001.

[231] D. Weaire and S. Hutzler. *Nonlinear phenomena in soap froth.* Physica A, **257**: 264–274, 1998.

[232] D. Weaire and S. Hutzler. *The Physics of Foam.* Oxford University Press, 1999.

[233] D. Weaire, S. Hutzler, S. Cox, N. Kern, M.D. Alonso, and W. Drenckhan. *The fluid dynamics of foams.* Journal of Physics: Condensed Matter, **15**: S65–S73, 2003.

[234] H. Wenzl. *Wasserstoff in Metallen. Herausragende Eigenschaften und Beispiele für deren Nutzung-Teil I.* Metall, **34**: 407–412, 1980.

[235] H. Wiehler, C. Körner, and R.F. Singer. *High pressure integral foam moulding of aluminium – process technology.* Advanced Engineering Materials, **10**: 171–178, 2008.

[236] E.J. Windhab. *Fluid immobilization – a structure-related key mechanism for the viscous flow behavior of concentrated suspension systems.* Applied Rheology, **10**: 134–144, 2000.

[237] H. Winterhager. *Über die Löslichkeit von Wasserstoff in Aluminium und Magnesium*, volume 12. Aluminium-Archiv Verlag Berlin, 2005.

[238] D.A. Wolf-Gladrow. *A lattice Boltzmann equation for diffusion.* Journal of Statistical Physics, **79**: 1023–1032, 1995.

[239] D.A. Wolf-Gladrow. Lattice-Gas Cellular Automata and Lattice Boltzmann Models: An Introduction. Springer, New York, 2000.

[240] S. Wolfram. *Universality and complexity in cellular automata.* Physica D, **10**: 1–35, 1984.

[241] J.T. Wood. *Production and application of continuously cast, foamed aluminum.* In J. Banhart and H. Eifert, editors, Proceedings of the Fraunhofer USA Metal Foam Symposium, Stanton, USA, pp. 31–36. MIT Press-Verlag, 1998.

Index

Printing: Krips bv, Meppel, The Netherlands
Binding: Stürtz, Würzburg, Germany